POEMS

EROS DESCENDING

EDWARD BUTSCHER

BOOKS BY EDWARD BUTSCHER

Sylvia Plath: Method and Madness
Faces on the Barroom Floor (novel)
Adelaide Crapsey
Conrad Aiken: Poet of White Horse Vale
Peter Wild

Poetry

Poems About Silence
Amagansett Cycle
The Child in the House

Edited

Sylvia Plath: The Woman and the Work

Library of Congress Control Number
2009913240

EROS DESCENDING

Poems by

Edward Butscher

A

The Amagansett Press

2010

A group of poems in this book was published as a chapbook by John Pierce's Dusty Dog Press (Zuni, New Mexico) in 1993: "Wake," "A Crooked Man," "Tartarus," "Exercise," "The Big Bang Theory," "Hog King of America, "I Do the Police…," "What Are Poems?" "Another Passage to India," "Barring Decapitations or Kind Embalmings of Sudden Acclaim," "Final Fuck," "Half Mast," and "No Man's Island."

"Ignatow Eats into the Sun" was written for *For David Ignatow: An Anthology*, compiled by Robert Long (Canio's Editions, 1994).

"Summer Composition," "Painter Man," and "The Times" were chosen by Richard Houff for *Scorched Hands: An Anthology of Verse and Rage*, published by Pariah Press in 1996.

"Dream in the Afternoon" is in *An Absence of Amy: Ten Years of the Amy Awards*, edited by Paula Trachtman (Amagansett Press, 2005).
"Ode on Cunnilingus" and "A Theory of Motion and Beauty for Wallace Stevens" are in *The Light of City and Sea: An Anthology of Suffolk County Poetry 2006* (Street Press, 2006), edited by Daniel Thomas Moran.

Other poems appeared in *The Amherst Review, Apostrophe, The Brooklyn Review, Confrontation, The Connecticut Poetry Review, Ginosko The Hat, Home Planet News, House Organ, Mobius, The Modern Review, Stray Dog*, and *The Tule Review*.

ISBN 0-943959-08-X

Published by The Amagansett Press
Drawer 1070, Amagansett, New York 11930
theamagansettpress.com

For Paula

and for Amy Elisabeth Rothholz

for always

PREFACE

The ensuing sequence, the last poems I will ever write, completes a circular voyage of discovery launched by the 1976 publication of *Poems About Silence*. Gonad-deep in a fierce midlife crisis, its struggle against the tightening garrote of Thanatos seeks to keep intact a frayed moral self without surrender to debasing mystical impulses, even as it nudges the obscene toward the sublime.

If not free of delusions, neurotic or aesthetic, *Eros Descending* at least remains true to its author's secular American humanism. Of course, other snares still lurk below decks, among them, old age nostalgia and a quaint faith in art's significance and permanence. But they were innate from the start, a conviction shared with Nietzsche that only "as an aesthetic product can the world be justified to all eternity."

A problem for any poet at an advanced stage in his or her career is the tendency to write despite absent inner imperatives. Though not that kind of poet after 1976, I now sense in myself a regressive urge to repeat what does not need repeating and become what Picasso warned against: my own connoisseur. Consequently, I feel compelled to echo a childhood hero's promise that "I will fight no more forever." (September 30, 1999)

Postscript: The appended notes have at their roots an early urge to preserve history as an extension of a psyche experienced by others. Besides giving blunt voice to Death's implacable power, they complement lyric reference points with the linear pleasures of prose and do not have be read, except by those, like myself, too obsessive to avoid doing so.

CONTENTS

I

OMEGA

"Secret grief is the cannibal of its own heart."

Herman Melville

And locked me in her heart, as one might lock
An old man, weeping, in a rusted cage.

Conrad Aiken

INTIMATIONS OF MORTALITY

(Flushing 1945)

I watch the enormous tulip tree age
outside the porch window, anchored
to the earth by exposed roots
watch its leaves dim, flare up
then brittle and darken
feel fingers and toes
curl in sympathy
watch them tumble
like charred parachutists
to an ivy-chained lawn

elephant flesh knobbed by cancer
growths like a wall's stillborn sons
wounds of too many winters

but the risen flock will not return
or turn again to reform the cloud
of its radiant eruption.

No basket baby every cried in its crab
arms or fled with wind-wailing seeds
from their reluctant release.

No young lover ever etched a chrysalis
heart in its stony flank or picked
scabs from its frozen skin
for sowing his ashes.

Naked now
               its palsied limbs
groaning in pain at every breath shift
the tree does not know the fact of death

no gambler having hanged himself
like strange fruit
from its reaches

but I wonder if a woman's paranoia
could dwell inside the scream
of its elongated bole?

I should inform you, and would inform it
if I could, if I could stir my self
from this wheelchair poem

that fall and death are not alternatives
to existence
                   that spring also croaks
its obscure marsh lies

that the feather of wild lips
across a small dry cheek
can gouge crystal scars.

There is no Frigate like a Book
To take us lands away.

Emily Dickinson

FIRST BOOK

Mrs. Beaver hung her laundry
on the fingers of a maple
trusting them not to stir
or let the sparrows
tug them free.

And Miss Robin Redbreast
paid a visit to pompous
Tom Turkey to lament
the bride price
of eggs.

Only Uncle Wiggily hid alone
under my bed, weeping
because he had grown
old and could not
escape Mr. Wolf.

BREAK FAST AUBADE

Comfort me with fat farm sausages
French toast and sunny, runny eggs
for the loss of that teenage appetite
that pried open every dawn surprise
with a dream-crammed cock!

BARRING DECAPITATIONS OR KIND
EMBALMINGS OF SUDDEN ACCLAIM

Tawdry figures silhouette
the horizon of unreason

hieroglyphs of lost learning
or much less than that:

amid RKO rays of sex and fame
George Sanders sank, committed
suicide out of sheer boredom
or so they say he said

    and baby-faced MacDonald
    benign as Emerson's marble
    dotage

        wandered off in quest
    of a ruined daughter down
    wrecked corridors of his
    elaborate mind before her
    absence gored and ate him.

Faulkner in his father's high
boots stumbled (sober) out of
the night's vixen accidents
to neigh, "I don't want to die."

    The body begins to intrude
    especially if neglected by pool
    abstractions beyond the glass
    or bootless kite shouts

to ache and contract alarmingly
sledding self toward mind's arctica
its sole relief a formica counter
in a corner candy store where
neon suns keep summer in flight

and the obese owner's butcher
apron conceals his manhood's
debit in cigar-stained airs
he will nurse on when wife
and loan sharks tree him.

Sheila was there as well
a steady flow of rubber swells
and chatter helping me serve
the orbit of neighbor faces
mooning our awkward dance.

Which way to go? *A puzzlement*
to ape phallic Yul's caked lungs
as Sister Teresa battens on
the black eggs her church bakes

Sheila of the coal curls
and precocious breasts
whose charm-capped nipples
trailed kisses across my
taut and breathless back

in the absolute opposite
of factories, sucking down
ideas and condom ideals
where a monster writhes

Sheila at a distance up
close, letting me worship
at the curb's chalk altar
watching her watching me
until the dust erased us

at the hard edge
of the hardest age

bolted school desk marooned
below Sheila's sisters bearing
maiden mounds of Hiawatha's
murmur to my secret ear
mothering imperious desire
in Michelangelo's lacuna
between agonized
hand and heart

the girls between sheets
like voluminous petticoats
of overturned dolls, peeled
down to a tingling syllable
actual flesh dense and hairy
as the texture of specimens

her mouths moist upon me
until mother and child
inflict domesticity's
maximum sentence

sad eyes betrayed by metal claws
that piston against air with frantic
lunges of Lucifer's frozen wings.

I FEAR THE DESCENT
OF A SHOTGUNNED CROW

Aspiring plums
hang slack
from my limbs.

Grey rats aloft
reiterate
cave reversals.

We cannot bear

higher web flights
sticky
with father honey

the further loss

of situations, wing
darts of anguished leaves
thrown down into darkness

faces that alter features

falling with him
as a foal
of the sun

furnaced alive.

Neutrino blinks
between being
and not

powerless words
like hands
over ears

we cannot comb
old screams from her skull
or escape the adhesive lair
between her scissor thighs.

PAINTER MAN

For Estelle and John Opper

When we talked, he and I
painter and poet, bleeding lilacs
over one another's shale profile
it was usually about imagination's
drive—generic as salt or milk—
to sift all that nature never spawned.

He had once hauled his family west
and honed Wyoming mountains
down to jagged human size
gifted by a dolphin-swift wife
willing to erase her Athenian face
with the grace of surrendering
what men can never truly yield
in their yen to sail the void
that empties a heart's hull.

When we talked, he and I
abstracting aesthetics into mirror
glaciers, cubing a Hampton sun
we never resolved where form
and orange diverge or dissolve

forgotten wives busy at their
still-life tapestries of explosive fruit
and fish ruins and florescent flowers.

EXERCISE

Aided by an ailing air-conditioner's
prevailing winds, rogue icebergs sail
by as I imitate a skier's extreme
unction, kilometers spinning a face
almost blind below brine droppings.

Raus! Raus! Raus!

Larger than life, I have mummified
under layer upon layer of fallow
years in this obese middle-age zone
like a West Coast redwood with biblical
roots, desperate to hide heart decay
inside orbits of rational decline
the sane telescoping of disdain.

"Good for your circulation!"
"Good for losing perilous pounds!"
"Good for huddling muscle masses!"

But goodness is easy to imitate
when ponderous after younger bodies
when larded by cellulite life-savers
caging one safe from shark erections.

Doubtless it is a Teutonic machine
though disguised as a Swiss neutral
intent upon redemption, exporting
missionary denials, contrary flows
where veal and astral mind copulate.

Raus! Raus! Raus!

Faster and faster, I league mountains
and let numbers reel time to a blur
when a cousin's gardenia bloomers
slid down to let frost-bitten fingers
stroke incredible balloons of flesh
that collapsed us into nervous giggles
behind the doilied sofa where aunts
sat to sip cappuccino and delicately nip
at the sugar rotting each female organ.

FIRST LESSON

Spain. Africa. Portugal.
Cities of gold.
Coast of ivory.
Cape of storms.

The geography of colors lay
jagged as Jello shards
under my knuckles.

I could not voyage to them
or from them, snagged
in the snarl
of their
names

ebony rivers snaking out of
reach with Sunday laces
refusing a bow

parenthetical mountain ranges
like a slave's pubic hair
worming among sheep

and tiny Italian ships impaled
on petrified crests
until my sweat
washed them
free.

I read the atlas backward
from a polar icecap
to the vast Sahara
in her foreign
glare.

All that tread
The globe are but a handful to the tribes
That slumber in its bosom.
William Cullen Bryant

WHEN THE LIVING IS EASY

Alone in the library simulating England
he crosses daybreak in my vinyl chair
a storm of broken atoms teeming through
glass to bite at bloodless ankles as he
imagines full-voiced organ stops bulging
dead veins that flatten out into a spot
less ocean sky beyond the French doors
where the sun shells a besieged city
spans an engorged Mississippi to firebomb
California's bleached pleasure domes—
sparks storming down to ignite wispy
beards of Japanese prisoners tending
their dwarf grandchildren and trees.

What shapes shape these puppet fingers
I sense in the lapsed hand that yearns
to regather sweaty breasts and thighs
of uniformed girls at play in a meadow
that never existed? To him, of course,
without cost, it matters little, if at all
eyes fisted shut to see what seams
have come undone in sleep, what further
erosions have harrowed a shaman mask
into the grotesque grin I will wear
when relentless leafy earth thumbs
me into its cryptic first folio.

ITALIAN POPS

Horace had hard Sabine hills
to hammer into his tombstone
Catullus teasing his fierce feline
erect, Virgil dancing naked
on Dante's suburban barbecue—
and Verdi's immortal arias
masturbated junkyard kids.

Artists all, we always yearn
for a corner candy store
where we can kiss the air
waked tremulously fair
by women scudding home in melon
swells and escalating mini skirts.

A THEORY OF MOTION AND BEAUTY
FOR WALLACE STEVENS

A sapling urge serpentining
between the shade talons
of a mothering elm

wrenches
its own elephantine image
from its own supple shell

as does the solitary lark
stitching up pockets of wind

and the monstrous sun
devouring its entrails again
and again with wounded fury.

Can the mover be moved
the theorist theoretical?

Angels, after all, are raked
from children's scalps each day
and settle in the lint heaven
of a Victorian hope chest.

The question fingers face value
but cannot elude time's charge
the relentless computer
tapping of gum-chewing cashiers:

an old moth hole in a yellowed
bridal gown looms the throat
of a tumbling raped girl—or not.

FIRST DEATH

I was ten or so
when the boy died
the stranger whom I knew.

He had been standing alone
on the porch of his mansion
when the pillar fell on him.

It caught his acne-harsh
forehead like an aimed ax
splitting open his temple.

At the funeral I wondered
what magic smoothed the dent
and what he had been thinking

about when the pole groaned
and collapsed his crew cut
like a circus tent.

Had he been thinking how rich
he was, his two-acre yard snug
around him as a cashmere glove

or had he been smelling
the lilac bush bending to him
with blue breasts and lights?

Perhaps neither. Perhaps he
had just picked his nose
and felt pure as a flute.

IGNATOW EATS THE SUNSET
For Old David Hero

No pie in the sky is wide enough
to circumscribe a poet's cold eye.

Squat in the doorway of a house
near the end of the island world
and catch each tumble of leaf
in hands cupped trembling to net
sudden up surges of winter light

solitary as an out-of-season
loon haunting an absent lake

another Utnapishtim sapped by
age and the unappeasable appetite
for injustice of a replaced wife
perched on your oarless shoulder
like Poe's mechanical parrot

although it is Horace you laud
sieving gold lines from stone.

For a moment, there, not here,
a thigh flex and breast thrust
launch the day into a fiery noon
before you resume the looming
of twilight motifs from knots
in your robin-speckled flesh.

Surreal events in simple words
is the dinner you serve yourself
that we sons of a fatherless god
may savor the deaths you rescue.

"I DO THE POLICE"

I said her nipples were scarlet
as dune berries dotting white
sand as she poured her
breasts into mine.

> *Grin awry*
> > *the protest artist*
> *totters at*
> > *the threshold of a café*
> *famed for its nautical motif.*

They weren't really, of course
more dun colored, burnished copper
under guppy mouths
of nascent appetite.

> *His umbrella wife collapses*
> *around*
> > *his bare ankles*
> *on the last step*
> > *weeping steel teeth.*

It doesn't matter, really
as she rainbows above me
arching us up
into the rain forest
where eternity has roots

where the sea never shells
the milk rage of a whale's
wall eye
when it blunders ashore.

FIRST LOVE

My hand gathers years
as I write this, clenching
itself into toothless gums.

The older girl loved my poem
but her bully boyfriend
ripped it into snow.

I fought back like Galahad
blood glinting off my fists
when they scraped cement

weeping that she should see *me*
vanquished at her feet, a wreck
of chaste intentions.

Breasts whisper now, nubile
student eying me brightly.
Too easy to pare her bare

of tissue blouse, tight jeans
to break open her ripe body
across the desk like a gift.

I could. I could (though sweat
freezes me in place) bring
her down like a tame fawn

but I cannot breathe
or fight free of the life
passing before my eyes.

Love is restraint, not ecstasy
probably always was, a holding
back of desire's iron hand.

TOWN & COUNTRY

Lush-shouldered dunes slouch toward
the rowdy sea, shrugging off
luxuriously porched cottages
behind a still mind
as I tally smeared casualties
to the tunes of Met static
and Golden Oldies.

Rag-tag flags of fallen comrades
(anarchists and Republicans
alike) littering the sun-stunned
highway mile after vague mile
attest to the miracle
of his escape.

Spun dizzy as an Amish weather vane
by Mercedes storms, then scurrying
between Viking and Cherokee raids
narrowly eluding a Japanese sub
before reaching the weed border
that had first caught
and released him

the squirrel celebrates his feat
in a warped side-view mirror
by lurching erect to sway
like a whorehouse drunk
tensed for another try
at going home.

NEW AGE

The pure plush breast of a vole
smeared by a ragged red smile
cradled in a quivering hand
gives Death a bad name.

FIRST SONNET

Alone with St. Mary's pews that December
while being punished for some mortal sin
(unrepented and now unremembered)
I saw a god's breath evaporate in
the gold tabernacle and heaven-bent spears
of wood and plaster clouds, my atomic eye
penetrating at once to the light year
truth of a toy town under immense skies
with needle fear and gush of joyful heat
knowing then that leaf-thin skin always curls
up to die in grandpa's blood-stiffened sheets
till blessed by a girl's undressed gift of pearl:

art's twin columns, caroling spheres of cream
a touch and taste of the sweet place between.

BIG BANG THEORY

Hard to credit: a rose pinprick
of energy pulsating whole empires
into black loneliness
                        when after before
a minute ball
is ignited
by a desire
so intense
it cannot stop exploding and expanding

though a cupped, coddled
and candied duffel bag
believes it truer than a starless womb
impossible to imagine
if not invent
                  a sailor's itch
scratched scab divine
                        by returns
to post-coital rage
and bareback rides

atavistic images
pounded on rocks and wrung
into a trickster mask

that ravages a woman in a rented room
who triggered the first dung fall
of grace into lesser beauty

her compliant dunes
her wild swan tremors
her wave-urgent breasts

nursing him into manhood
and its shell-hard hatred
of the pliant flesh
bearing him there

then a killer's lobster retreat
from dried complexities of dawn
sheets and the Other's tainted body
scrawled stomach scar and shabby
wallpaper hounds hurrying him
down strange stairs.

That crime: that debt: a complicity
of motive more convoluted than axon
firings into punctured cobwebs

sent the silver bullet soaring
across enemy lines
and through the nightmare
empyrean of symphonic spheres

fins erect
   gills gone
      tail aflame

for an assault upon
absence without end

afraid the thrust might disintegrate
from sheer puniness
or melt and crawl

brought to howling knees like dinosaur
hordes under a judgment of stones
or a sissy who had forsaken
his father's ferocious sky

that hopeless journey backwards
against the pearled wall
that nurtured it
intact and apart

forever apart.

CHINESE PROVERB

I remember the sad one
about song birds
making a poor feast
but can think now only
of the shy-eyed student
in her starched white blouse
who barnacled my bare
arm with almond breasts
at every shocking stride
and sudden green upsurge
that always graduating day.

FIRST LATE SHOW

Snug between the toes
of a hunched piano
stationary
to raging house rales.

Get out of town
on the noon stage.

It is lost
leaving behind
dirk-clean moments
of trust and truth
       slipping in
slipping away.

You don't care.
You never loved me.

Cracked round thighs
of ceramic Greeks
long since dead
their robes like rills
leaping bone-white
       from lean hips.

This is it, men
synchronize watches.

The clouded bulb
       of a breast
hung (eclipsed)
       over all
too pale and smooth
for twilight kisses

Don't cry, lad
she's bound to return.

A grandmother's life
       snow-flaking
from a slumped sleep
       into pauper fingers
and my granite eyes.

THE TIMES

She lies in the street absurdly
still: floral kerchief and well
tailored coat unwrinkled and
unstained: smart pocketbook
shut in her unsurprised gloved
hand, only the bag of groceries
scattered near the other hand
limns the prankish violence
that drops her at our door
to be sniped at again

HOG KING OF AMERICA

Born in steel cubicles inside a simulated
barn (sanitary as flossing after dessert)
the hogs are shifted to another series
of sterile ice-tray cells when ready for
serious weight gain (scientific as radio
crystals), then shipped at 220 pounds
to factories where workers and machines
blink in digital unison to drain their
dreamless lives of all VCR reruns.

Mixing memory with desire, which Blake
preached was electric and Freud rewired
can we (like Proust) clot the polluted
flux of consciousness by freezing it
into art as it pours back into the black
aorta flood that arked us to this amoral
moment and means, or must we be as sick
as he, as gnarled by Europe's perversions?

Perhaps an answer rides Jefferson's West
with the Hog King and his broken rough
product, his acute entrepreneurial grasp
of Nature's spine cock, cold mechanical
reapings of chained prey that never taste
the ecstasies of hedonistic wallowings
or squeaked protests against the razor
precisely penciled across bared throats
a rutting in sawdust where gods fuck.

The pornographic photo that stuns the sun
is Beauty and the Beast: a smiling young
woman in white intern jacket that clips
her breasts, spiral pad and ballpoint pen
poised to calculate the Hog King's dis
tending profit margin with educated de
tachment from the creatures crated behind
aluminum sidings in democratic isolation:
heaven and hell harnessed together at last.

Reflections of vacant eyes and the genius
that renders them from essence lay siege
to the island of lotus-eating swine our
kind tends to fantasize as Eden's clime.

Too old to remember, too young to forget
I cannot abide the chill lack of victims.

I wish I belonged to another species or
another era when sledges of death were
usually a surprise—and cruelty mere reflex.

FIRST DATE

Standing at the edge of the vacant lot
where a baroque movie house once stood
stiff as a comic sentry left behind
by a silent war
                         I do not mind
the bay wind stabbing me in the back
or the cold-eyed prods of strangers
pausing to measure the menace
of my senseless act.

                         Only the debris
disturbs me, wounds me, twists a bayonet
deep in my bowels, dumb heaps of plaster
columns from ivy-painted temples (crown
less in the dust) and sham-Roman arches
hammered by real time into coffin bricks

the mute artifacts of serial adventures
reduced to rubble for rats to disgrace.

I held my distant cousin's breast tight
in the balcony's star-salted blackness
the first naked breast I had dared milk
worshipping at the squareness of its
miracle nipple as the model circus train
groaned, arcing sparks, through the rain
then crashed, crumbling like a necklace.

Metal shrieks and lightning tore animals
from their cages, played over our locked
figures, my hand trembling from the task
of maintaining its passionate delight.

           "I love you,"
she moaned expressively above the noise
nibbling at my ear with kitten-tiny teeth
shifting with the earth to a fuller bloom
far too heavy for me to ever bear.

APOLLO 1

I refuse to compose Japanese gardens
parchment skies, pen-fine peaks
the gentle raven scratchings
at cattle like ripples
slowly slipping
out of time
>the order of white pebbles
>frozen into fangless adders
>ornamental trees and bridges
>so tiny their mirror ponds
>cradle an egg-whole world.

But I can heft the womb weight
of a burden slung from poles
lifeless and lacking blood
(a tapestry's myth malice)
>and I can survive
>inside an art that
>embalms breath
>by severing heart:
a lean man and his elm oar
sinking in a seasonal tide
that never stirs or sings
of victories over walls:

fishermen drawing nets
from perfected clouds:
>women and leafless boughs
>waiting for warrior samaras
>to return and blossom in them.

The sly laughter of frost tears
mocks an elemental wind.

THE 200 YEAR OLD TORTOISE
AT THE BRONX ZOO

They look. They look and laugh
but her pitiless stare does not leap

lacks the human weight of dancing
tiaras, an erne's pinpoint hunger
measuring the ocean's quick hares.

The tortoise boulder will not move
wills, instead, her pious watchers
to their scripture dooms, a loss

of primal earth, whirlpool screams
snatched from ritual rape scenes
the mirror's covert silver deities

gurgled into unseen flames, curled
ghosts, billowed sheets of ash skin
flapping at her wire stalag, her air.

As if to speak, those sickle jaws
hinge wide and leak green phonemes
ululating without sound or sense.

They spin. They all spin inside
the diamond pivot of her blind eyes.

...sexual love is but a mad craving
for something which escapes us.

Michel de Montaigne

ANOTHER PASSAGE TO INDIA

I

Veins and stomach fangs
conspire to stake out a failing
body's frayed boundaries
the heart's sudden rampage
down dusk's gaudy mouth.

I taste the chalk the dead
leave behind when arteries
sputter and clog with too
many snarled memories
ways of not dying.

Let me daydream instead
of an infinitely delicate tongue
busy at my distended penis
and of the bat woman's attic
thighs bulging camellia cushions
for my sunburst risings.

One huge shadeless light bulb
haloing three tarnished Becketts
pins me to this bed, this room
this mother-and-angel pair
of altered cats, sprawled atop
floating spines without a wish.

I tease open the silkworm fissure
that unfurls Nana's leaf nest
where I tenderly unstitched
a baby baseball's waxed wings
before bashing in greenhouse glass
where I always fall alone.

II

"Tell me the way. That's not the way."

In the village where she was born
daughters remain a burden
however beautiful, blighted
seeds to be thrust
inside hoarded manure piles
until a monsoon dislodges them
or a teenage husband's torch lust
until smeared on squamous walls
and used to fertilize concrete
soil, American ferry rides.

I kiss her recent breasts

one by one

then shift, gently
a mole hunger into a vortex
of cobra amaze
                to ape
more precise nipples
in an endless runt quest
for what she it they
could not possess

                    jaws askew
with a jade monkey's smile.

The hand at rest around her
waist slithers down to cup
a small dense buttock
straining for adult intelligence
crafty middle finger
                    crooked
to hook and enter
the denim nook
                    where natives
hide love's fearful
first quiverings.

                    III
These are monsters that I skein
from the unspliced fibers of a brain's
moist maze, mistakes of focus
that achieve what learning loses:
howled writhings in a blind alley
or a crib, a stranger furred like
your father, decayed animals.

Nothing is rehearsed
everything posed:
                    an obese guru's
candy-striped arm noosing
a slender, long-haired girl, unruly
erection sheared off by her
twin holding a camera
in India
                    where cows amble free
udders shriveled into stones
to ballast Ghangi corpses.

"Certain sects worship
certain parts of the female body."

The prisoner's opulent lips shape
the words below olive black eyes
that reverse every slow gesture:
seduction and reluctant desire
knowledge's bittersweet grasp
of all graven images, our
cameo silence saved by
an eclipsing wink

So do I! So do I!

II

BETWEEN THE ACTS

"Why should a dog, a horse, a rat have life,
and thou no breath at all? Thou'lt come no more.
Never, never, never, never, never!"

King Lear

...our existence is but a brief crack of
light between two eternities of darkness.

Vladimir Nabokov

SWEET WILLIAM'S CONCEIT

To abstract an abstraction
(Emerson's "fossil poetry")

   love is a verb, then a pronoun
   the space of deep space between

you must extract the leech coiled
inside all cunts and icy-eyed cocks

   (a flame and its fear-fevered
   wick painting a SS basement)

to feel the full organ weight
of a rose's fated fall.

Two sweaty, writhing human forms
jig-sawed into a single self

   ape the orphan of their grapple
   for adjectives to freeze passion

shaped from glistening body fluids
that smear separating cell walls

   into the lightning-tortured wooden O
   where an old man bellows his loss:

a lashing lasso of words to string
up the universe's death sentence.

THE REST IS SILENCE

Heart suddenly an ice-encased moth
I am stalled in this hylozoic room
that will outlive my panic attack
and the artful hierarchies of texts
and artifacts I amass against its
terrifyingly tall indifference.

I can't invent an absence as dense
as the collapse of a megaton star
eons before Nana's donkey sang her
into *Bella* like a Medici princess
snug in this white-whiskered black
chair (once clawed by a one-eyed
alley cat) so black not even a pin
prick light can pierce its polished
hide sag or needle me whole again.

The ontological ignites with an O
of anguish and gas a baby bubbles
nothing more or less awful—unlike
the phenomenal nominal of *nothing*
that slaps into existence carnivals
of comet faces and sea sounds arcing
a life toward the no thing it names
the nouns of *mother* and *father* sent
tumbling by exploding gerunds above
the round horizon wheeling creation
from bed to bed, house to house
as love's mimicked lexicon mocks
the scowl of a woman who keeps me
locked in bed the entire six months
of an unforgiving first grade in fear
of the Nazi air her breath infects.

I rise and retire each day tasting
the dry turd of my own death behind
second eyes monitoring every action
for signs of the disease that will
leave bits of hair and bone in a yard
where a fire blackens the earth hard
as these abandoned highway walls.

De Kooning's clown colors and threads
from his last will-o-the-wisp decade
fade into tattered pastel ghosts
of an evanescing awareness.

When a fat one-eyed mother cat falls
or an actress daughter's geyser laugh

when a father is formalized in his
only suit by waves of white satin

when a strange woman falls and folds
into a painted sea of bathroom tiles

when a heart heaves up what a mind
must swallow of a sweet aunt's sense
less sentences each visit and night
to the "inner room" of walled peace

then neither will nor intention can
anchor intangible consciousness.

Unhorned and left to rot for horny
Asians, the black rhinos dwindle
day by day on the kitchen calendar
where an ape apes Rodin's thinker.

The local poet and former professor
(now age 87 but still dapper in his
charming coldness and beeper smile)
reads under a savage afternoon sun
at the college that had exiled him
to a band of retirees and colleagues
that float, boated, above the flotsam
of third world and minority students
forced to attend and pretend interest
in an old white man speaking tongues
before escaping without a word when
the bell saves them from a lyric con
of an Hispanic housepainter's soul.

Neatness counts as a way of retaining
a universe of bed, book, and tomorrow
galaxy after galaxy aligned by will
to vacuum up the dust balls of dead
planets I find in every granny corner
satisfying rag sweeps at a gleaming
wood surfaces where the globe my aunt
bought me projects colors so enamel
intense they dazzle the dark passage
way from living room to toilet.

A Tuesday night fight on the Mohegan
reservation lights a Dancing Christian
taking a split decision, red and black
enlaced like a Chinese bridal banquet.

Blind as Beethoven scoping a piano's
braille pulse with a bleeding eardrum
inexorable as the deaf Black Plague's
squeaking dog carts of flour-bag bodies
the blackness that squeezed Donne free
of his wife's silk winding sheet back

between choir legs of Mother Church
on the silkier sway of a new wife's cum
mauls the loose seams of my sleep.

Even at Nana's baroque Sunday dinners
that plugged up eternity's suction
with the clatter of spaghetti-piled
plates and steaming heaps of chatter
I spied the skull gleaming in at us
and savored the acute oregano irony
of her daughters debating a grave's
ideal site over a river, below wind.

Hidden girders and high cement ribs
of my apartment support the entire
three-winged building with vaulting
conceits and Kant's categorical be
coming godhead as I wish for the joy
of being a Turing toy without being.

If Braque could CATscan the skeleton
of the universe through a fly's eye
without leaving his cluttered atelier
(what a mirror reflects when dropped
   by a crazy lady's unclenching fist)

if Cornell could cushion infinity's
terror with a Flushing-hewn coffin
wrapped in his mother's thorny arms

if Perchik can mount a pyramid lament
to his looking-glass sister's polaris
with toppled Jewish cemetery stones

if a crumpled spider body can crawl
up a telescope of mathematical ego
to the ragged, bleeding lip of space

if I can feel a one-eyed cat asleep
beside me like an amputated leg

can I ever ever really really die?

Art is the dark wish of all things.

Rainer Maria Rilke

DUST TO DUST

Looking under the bed was always
childhood's frantic morning ritual
searching for the cockroach horde
that shunned the vast linoleum tundra
of a renovated Smart Street apartment
where an inebriated Prometheus labored
to lure in tenants to defray fire damages.

Dust under the bed in borrowed houses
when I roomed beside Grandpa drew
Saturday's women, helmeted in scarves
and wielding long, rag-hooded brooms
with the nesting efficiency of the female
though Grandpa's caked floor of blood
would not yield to their mine sweeps
leaving behind wispy planet husks.

I rescue an early poem from my first den
seven stories high, its bedroom windows
slashed by splashes of pirate-black paint
and the whiskey laughter of truant wives.
Fluff remnants of Shakespeare and Freud
intermingle here with Visigoth hoodlums
swarming over Rome's crumbling walls.

Let us escape to recollect autumn leaves
parting underfoot, raked into gaudy piles
of a tower poet's collapsed angel wings
and burnt for neighborhood incense:

Phoenix ash compacted in stadium turf
that does not tear under falling bodies.

Her lips refeather an old man's flesh.

III

ALPHA

"'Stand round me, men. Ye see an old man
cut down to the stump; leaning on a shivered
lance; propped upon on a lonely foot. 'Tis
Ahab—his body's part; but Ahab's soul's a
centipede.'"

Moby-Dick

WHAT ARE POEMS?

The grunt of a lamb girl
when I ram myself home
sag a made bed

the way a mother sways
when a sheet is lifted
gently replaced.

There has to be a third
stanza to rescue me
from the others.

I saw a staring virgin stand
Where holy Dionysus died.

William Butler Yeats

FROM SHORE TO SHINING SHORE

Dear Lucy of elevated Queens
as the year of the Golden Dragon
spurs its Korean and Chinese wives
scrambling (spread-eagle spider-wise)
to become pregnant with purse excess
let me pen you an authentic love poem
from the Roman outpost at Island's end
where investment bankers hire pleasure
domes and lawns of Hispanic gnomes:

> *to whet and wet*
> *to bait and bite*
> *the sad sea mouse*
> *between swan pets.*

So you can see (or siren, if you will)
an amorous lyric remains the roughest
to ascend without sacrificing shark
attacks and tragic undertows, tales
of dead children symphonically scored
Africans burnt alive inside a church
baking earth brownie-dry with god wails.
My precocious goddess avatar

> *metamorphose*
> *the furry worm*
> *that flies to rot*
> *Blake's factory rose.*

Every Aphrodite has her Aphroditos
a slime-lime snake in an enamel apple
bearded white by an ancient appetite
that bibles his beast roar for nymph
meat, my awkward sideways shuffle
to embrace leaps of your iridescent
laugh where loss first rainbowed wish.

ODE ON CUNNILINGUS
(The Road Less Traveled)

A tongue from brow to ear
feathers, dips, dives in
then descends corn-haired
breath to brimming lips
engorging a mirror fluid
mouth (muscling eel feel
grappling for root ends)
teased nape and neck next
swooning with Ovid's plush
swan, undulating reversals
under the Ur-exiled artist's
softly reiterated strikes
until the flushed fleshy
globes are slowly stropped
to a stuffing firmness
each delicious nipple
moistened shiny as button
candy by need's mindless
separation from the heart
down to the downed belly
slyly winking rue at ever
having left home to lather
the small crest of a furry
burial mound for soldiers
unborn before the sleigh
ride rush through a copse
of tumescent, dream-blurred
vines, sniffing strangeness
of ocean wreckage while
kissing open salty lips
of song to give her primal
pearl a tongue and a voice.

WAKE

Creation has an hour

of winter dread
that stalls the heart
just before numb hands
can net its swift fin.

Huge black bulls bellow
and hulk on sapling legs
when nude boys and girls
climb down from their
laurel crosses to dance
a cave into fireflies.

Here: rain-bearded dawn
(grey as spent cement)
blots a motel's windows
with squat mirror tears
until nothing is clear
but overturned stairs

and the moon rock behind
our deliberate blindness
is smeared bright orange
with old offal and blood
signs of a denless beast
that devours its children.

I conjure a puma nesting
inside my ribs, breathe
a panic of wings wheeled
from an owl's oak crotch
while reassembled thighs
descend into marble.

TREATISE ON LOVE AND BEAUTY

Aunt Ida's fall into woolly Hellespont
unfolds with a disobedient teenager's
unpainted dive into an old gay poet's
moon-scanned wrinkles of loneliness
like a sinless son flung from a tower
to wail without words into literature.

Warriors on shields and shaggy bulls
that stampede in savage waves to smash
asunder a neglected wife's conch heart
echoing her monster child's death cry
strangled by the hands of a foreigner
and a wicked sister's sea-scummed web

are among the images of myth quickening
her girl-like gestures of popcorn joy
and charming plaints about a character
fate of split ends and family clashes

our fingers unwinding and entwining
across a rickety desk, like wrestlers
in a naked Greco contest slow-dancing
toward the circle rim of infernal sex.

I have to bang, hump, fuck, pound
alabaster bones into oyster insides
fear and lust eating out your cunt!

Her long hair, swept back in a silk
cascade I yearn to caress and nuzzle
can relapse at any time into Veronica
Lake's naughty swoop of indifference
to a hungry camera measuring how far
her full-lipped smirk would descend

her high breasts breathe under girder
spans of modern design with mounting
passion for what can only be imagined
strain to expose their buried berries
and the pits that ripen the nightmare
leers of strangers masked against her.

I adore the mind behind her indulgent
mother smile, more lithe than Daphne's
sly bird flight into a river father's
laurel arms when Turks rioted ashore

bayonets or short swords hacking at
a forest of women and children too
rooted to escape our dream of them
old men slaughtered like auk armies.

*I want to copulate and fornicate
and penetrate with a rapist's
fist to the grit bottom of you.*

Iphegenia of Corona, now cavorting
around a Maypole Mentor, whose
father bled her for the winds off
the isle that twisted him crooked
as the clan that had uprooted him
sly seducers lurking everywhere:

it is Longinus I should teach her
to sublimate the cypress shivers.
I have to tease her into laughter
and love blind to my thinning self
the sublime transcendence art can
insinuate between reach and apple.

After an Athenian mastery of stage
and *scienta* and human-sized deities
done in marble with an ideal grace
came the Roman drapery of the real

to realize her perfect lily beauty
grafted on a vine of imperfections
small enough to escape Hera's fury
and her brother's Picasso stare:

I will the fresh fruits and juices
of your body to salvage my nights
as you throb a wet dream's deluge.

her jaw's sidewise chuckles of wry
its tectonic plate easily unmoored
when I twit her about her boy-eating
ways, the belly button a Latina mother
had ritually wrapped into a turret slit

a scar splashed near a racy pulse
like a clump of coagulated cream I
cannot resist thumbing for a trace
of the holy hill nest where Psyche
drooled the hot oil of her female
thirst for a cave mystery's end

a birthmark seared into a slender
white thigh like the devil's witch
sign, a patch of grave-misshapen
mud she will not disguise or hide

that makes unique what seems mirror
weary in the breathless moment my
poem shafts her striding toward me
in space shoes and mini-skirt glee.

I pray to pedestal you gently bent
in a plasma robe of lipless globes
and hairless clit above my chisel.

This is what I should have preached
when she perched in my tiger-shaded
flat, sipping rye from a wine glass
like an adolescent Laura or Beatrice
when I ached to kneel and be cradled
thanking me for a lovely afternoon:

once I hugged a tomorrow too (like
you) and still embrace some Eternal
Present, but artistry courts truth
not frenzied couplings on its altar
of luminous crossed thighs, the eye
of a rooster, not the owl eloquence

of Abelard's rhetoric, Aunt Olga's
caged rage, not Aunt Ida's benthic
drift, to burnish the chalice you lift
to thwart descending blades of sun

as I brush dewy breath between your
sprung legs with imagination's tongue
as I gild the jungle of sweetest decay.

Petals on a wet black bough.

Ezra Pound

ENGROSSING DEATH

The moon
  soused Li Po stalks
  until he drowns in her

the moon
  a jewel the Bard
  pins on cuckolding night

the moon
  manic Plath sucks bald
  as her mother's bound breasts

the moon
  our monsters walk
  staining her in snail waste

the moon
  cannoning the window
  Olga flags with cloth howls

the moon
  I bury alive
  in my mind's every breath.

"…whenever it is a damp, drizzling
November in my soul."
                                        Moby-Dick

NOVEMBER'S FABULOUS ISLAND GIRL

I

An ear dangles from a fig limb
below the impassive cloud face
raked by rioting black birds.

Death's antidote
                    the Tahiti dream
is the sole dose of god aging allows.

It is 4 a. m. always,
                    and I'm terribly awake
twisting convict sheets into sleep
and another dream of a dusky girl
emerging from the gongless sea:

gorgeous sarong twirled around a slim waist
to burgeon modest breasts into glossy fruit
under my open mouth (against laws of space)
sweet nipples swollen to sting parched lips
like cherries still moist from Nana's icebox:

her incredible smile a swirl of ripe pulp
too scarlet for reality's reminiscence

her fuzzy vulva thrusting voraciously at my
half-limp penis as it rises in adolescent glee
and enters the mud-soft pith of wordless time.

II

Eternally now, the mad woman in the attic
of Booth Memorial's seventh floor, where
the poor are tended by imported nurses
Aunt Olga claws at shriveled dugs in quest
of the lush bosom that drove Corona wild

the most creative of the Cutillo sisters
with her ear for music and melodramas
who posed once in an arch of proud breast
under the undone halo of innocence that
soiled lungs and mind into windless sails

far from Gauguin's island of massive
native women at play in beds of sand.

III

I am a virgin, she apologizes queenly
but knowing how to ease me inside her
cradle hands mothering taut buttocks
salt lips at work on cheek and nape.

IV

The dream saves Aunt Ida from her
bay rock for another command post
with the Red Cross disaster unit
even as she lies here helpless

half outside a disheveled housecoat
maiden bones melting into crib cries.

V

Love redresses dawn's vicious light.

CROOKED MAN

When everything else fails, and sleep does
not seep from fossil extremities
to etherize pulse and mind
it is time to elide to Dad's 1940's face
with grandpa's silent straight razor

and dream a circle of kindergarten nymphets
chanting their song of all falling down
(flying asunder like Florentine doors)
to charm them with a resurrection
they can never forget or flee

to remount a grave hallway's iron skirt
and spy upon St. Mary's girls loosening
their regulation blues, letting white
underpants parachute to dust as they
sneak smokes and squat like piglets
over whiter bowls, gushes
of naughty laughter splashing
high mirrors blind

to enter a packed subway car and study
the graffiti *Times,* orca blunt
erection nosing the backward
swirl of a steno who cannot shift
far enough away to void the next
shuddering stop that wets us both
when I reach the want ads

to mythify parks, Central and Flushing
beloved Memorial Field, where a grass
ocean lapping at a concrete wall once
washed up two bodies, locked in spasms
of butterfly debris, the cheerleader

I idolized from sidewall eyes
and a collapsed football star.

Impossible to explain how her China
form could bear his dead weight
as they rocked beneath waves
of lava heat
                or how green stains
are ever scrubbed clean.

WET DREAM

When I lie down to sleep
without a soul to keep
Death straddles my ribs
like a boyhood bully
until I cannot breathe.

Penknife at a gourd throat
only hairs past puberty
it was much easier then
to spit out his blood threat
and humiliating terms.

Now I groan and embrace
a pillow's fickle grace
as the faces of my dead
slip from arthritic fingers
to rattle down eternity

Erato my only craft.

Beauty is momentary in the mind—
The fitful tracing of a portal;
But in the flesh it is immortal.

Wallace Stevens

OBJECTS OF DESIRE

I dream the dream of the Vietnamese
girl dying with an *haie* cry under me:

cans of Adorn (purple as Naples grapes)
and Woolworth mirrors that crack up at
each new wrinkle and a Westclox clock
now guaranteed to cup thirty-six hours
in a single sweep and knee-high nylons
opaque as tank webbing to snare pared
calves and lipsticks of fire and ice
and ice-cream crackers two for a dollar
week after dutiful week toeing a hallway
where discarded ladies sprawl, adrift
in the carnage of their broken parts.

Space *is* always anxious around an idea
without an object void to limn its loss.

An Italian fisherman, black as charred
wood from Nana's deconstructed parlor,
guards the godless corner where two
authors hang in their bourgeois smiles
heroic as a toreador in his net held
pose beyond my bed, at times, behind
flowing away in the waves of electric
lacquered hair I caress with Dad's
frost-calloused, fare-thee-well fingers

death terror at rest against a girl's
small firm body as it responds with
refugee despair to pirate incursions
by not responding, tiny breasts tidal
in their unexpectedly generous swell
below the caked salt of all yearning.

TARTARUS

Love   light   darkness

return to twist in air
that is not yet there

when the father sheds
keys   wallet   suit
of ordinary scales

pearly skull of stone
above the androgynous
sponge issuing a long
vowel sounding itself

when the father crawls
under his daughter's
garden canopy

cold as a grotto echo

laying his secret bulk
on her until she can
not remember or say
who she is and why.

There is nothing
he cannot do

when he lifts and balls
her Disney nightgown
into clumps of clay

when mouth suckles mouth
trickling white blood

nibbles through cloth
to lick tiny nipples hard
as Christmas candy

his lullaby arms
tightening her
into him

his instrument.

I am guilty as he
letting outrage
dance on the head
passive as she

of my own swollen
words made flesh
when he rubs her

gently   sweetly   cruelly

between unfolding legs
until the downy mist
of her yields a flood
scream that eases

us inside.

A DREAM WITHIN A DREAM

The shirtless old homeless man
walks a midnight Main Street
with a Raggedy Ann doll
crushed under each arm
a leer like a World War II siren
warping his unshaven face.

Nights earlier, while still
wading down Amagansett dunes
I had dreamed about St. Michael's
school yard (its high fence
inexplicably invisible)
coregent lines of children
clutching lunch pails and books
to their chests like anchors
or rescued talismans
as stately nuns moved
with mourner slowness
among them.

I tug the dog to silence
and watch the comic figure
disappear like matchsticks
into a pyre of street lamps
before returning to a dead
cigarette and reveries
of summer escapes.

I saw myself there, cap askew
brown bag bulging with Nana's
meatball sandwiches, standing
in a brother's shoeless
galoshes to wave at me
(a stranger) as I passed
until the pathos dove
deep enough to shake
a heart into rags
                    until I awoke, cradling
                    myself like a drowned son.

FREUDIAN MAN

They come to me, bend to me with their
piles of piping-hot teenage nightmares
in a bookroom made hazy by second-hand
smoke and my deceptively youngish form
dumping them in heaps of dismembered
tropes on the unsteady standard desk.

I sort and decipher them with a teacher's
care not to grapple free their bottom fears
or wishes too vicious to be long endured
knifing aside rough beast slouches for
a sunshine urge to harvest a healthier
ego from childhood's gingerbread house
and swallowing whatever suicidal dives
seem destined or desired by eel appetites.

She came also, Diana's dark twin, exiled
from Guyana, too beautiful in blackly
exotic cheekbones that recarved upper
class profiles astride a Keatsian urn,
classic features sanded free of acne
to excavate a princess smile sensitive
to scars on others, herself slashed early
by a rape—if too narcissistic for altar
fires, for sackcloth or kitchen ashes.

I could (and did) love her differently
from her classmates, secretly caught
in her lustily unleashed raven tresses
(her voice an insistent covert hand
massaging me near the adolescent
climax that would shatter her face)

but her stick-limbed leanness shivered
me into gyrating recalls of a mother's
gaunt depression (though startled once
when instinctively slapping chalk off
taut slacks by the two planetary swells
of flesh more abundant than imagined)
as did her need to flourish elsewhere—
beyond barbed-wire boundaries of art.

Her absence, of course, and the waste
of loving what now seems exotic erotica
haunt me in the silence between bells.
I recollect the moment of revealing—at
the throne pith of her most disturbing
dream—who Mephistopheles really was
and who his archetypal Faustus had to be
beckoning her into a Hades where naked
strangers clawed shadows from her body
to harden nipples high and low vulva wet

displacing his bearded grin between
her pliers-easy legs as he crouched
(ponderous and pedantically intent)
below a winged Victorian cape.

THREE O'CLOCK SCHOLAR

What we all need, my loves and I
from time to Drake Cake time
is a psychotic break, enjambing
a Honolulu sunset so garish it
blasts us out of the sluggish
curbside flow of flotsam routines

like an eruption of teenage acne
or a covert lottery infatuation
like a slippery porpoise penis
springing off an unmade daybed
after a class memory of a pair
of cream perfect breasts
and a Greek girl's sly
downward smile below
my even slyer lecture
on old J. D. Salinger's
incestuous pedophilia

and multiple fantasies
of her rising out of her
tank top and baggy jeans
to soar into startled arms
with a hydra-tongued kiss
and fierce hip thrusts that
arc us through the ozone

until our Argo explodes and cracks
apart, like any other old man's nap
of trumpet farts and false starts.

SUMMER COMPOSITION

Lolling in long sun-plashed afternoons
the language of it truer than the life
led above a seashore's halcyon humors

where I wrestle a beached body
near the tanned athlete
who scorned the death
each conquest scored

"I'm a good Moslem girl," she crooned
half-laughing, as we showered together
in the milky window light of Queens
after my dives for the sixty-nine pearls
of her bottom-fishing wit and wish.

lingering over a cigarette and coffee
on the side porch where ancient hews
and webs replicate a forest primeval

I read of young Bosnian women raped
by former friends and neighbors
with bayonet stabs at a tribal replay
of that original deflowering.

From school to school, the season had
always been an electrical sail over
hedonistic outfields and hard labor
on bicycles and in dank basements
rowings from bar stool to bed
until marriage remapped
habit's grassy harbors.

Let me count the days in languidly
sultry nights when I love the ways
left behind with the trash I now
tend and once again pin down
the giggling village girl
who rides a limp penis
into my killing field.

Hast thou named all the birds without a gun?
Loved the wood-rose, and left it on its stalk?

Ralph Waldo Emerson

MOLE MAN

Walking beside the raised earth tunnels
embossing the front lawn's battlefield
he pronounced them "hard," thus "old,"
no longer home to lusty moles blindly
humping their way toward tasty grubs.

He could not stop smiling or talking
as we reached the single earth snake
uncoiled under a sweep of willow tips
about raccoons (creatures he doted on)
wreckers of a hated neighbor's attic
before abandoning their alien nest
for more homey ash cans and alleys
where he set his toothless traps

telling me he could put a carton
of their kittens on a back porch
and they'd be gone by moonrise
in their mother's clever arms.

Memory conspired with his sentiment
to repopulate a deserted apartment
with that nude Gauguin girl stuck
fast as spider prey in a gold rug's
quicksand rays, full flush of lips
more aware than his as a firm hand
fondled a useless sack of testicles

pendulumed below orthodox poses
her self-reliant tongue relentless
in its quest for ancient cave fruit.

Around the house we slowly strolled
like scientists on Bikini's wasteland
until convinced that a single "run"
remained active, hardly worth the toll
of an ambush that required by law his
daily attention, adding to my urban-
lean store of animal and human lore:
tales of a Lily Pond estate (ghosted
at the time by divorce) where his "cute"
bandits capered for months on antique
furniture at the cost of a poor family's
annual income and of a feral cat he had
caged at a caretaker's request that ARF
refused to accept, the blue-haired lady
there petting it through bars to prove it
domesticated after he had threatened
to blow out its brains in the street

returning it, instead, to the same area
where he had trapped it, and his fee as
well, laughing out loud at himself.

On a daybed scene woven by a tacky
spread simulating dense marigold fields
the olive girl—more thoughtful than
her grin intimated and as loquacious
as he in our sweaty aftermath—loomed
an odyssey graced with absent parents
and a preferred sister, who ruled and
ruined her American design (sleeping
alone in a gauzy gown below the moon
held aloft as a mirror for her flaws)

while messaging ridges of a slack
penis with preoccupied affection as
I tongued thimble nipples into stars.

Yes, I nodded breathing her air, *yes*
assenting to both of them, delighted
that he would not trap the solitary
mole (actually a shrew, we agreed, I
had unhoused by accident), handing
him twenty dollars for a lesson in
nature, I was content to remount
a veteran forefinger as a substitute
in my wild gallop across dewy grass
and down Alice's looking-glass hole

achieving the sole climax I can still
ascend in this life of illicit pleasure—
measured without meaning or moral.

Tecum luders sicut ipsa posse
Et tristis animi levare curas!

Catullus

PIN UP

A side front tooth that tilts against
the others threatening frost spillage
Roman ruins set above attic symmetry
lean legs fleshy in the thigh straddle
the chair I yearn to be sleighing over
into the clothespin burst of a flushed
thrush, ascending to the sun of a smile
so aware it mothers crib inclinations
between fan-delicate ears cupped like
elfin caps to radar rumors of Ecuador
atop poised breasts deemed too small
by childhood bullies but now perfect
in their tented nipple sway of tulip
thrust and swell as firm as bun-warm
avocados to embraces of lip and heart
anchoring a flat stomach's unlocked
wink at the tongue of a naked desire
to rise and kiss the crooked trail
parting your unpinned surges of hair
scented by crushed jungle garlands
and cocked politely, as if listening
to the way my lecture on Latin lyrics
and the decline of Western culture
drifts off into a humming-bird blur
of dwindling content you laughingly
mock as I cure my hands at the flare
spasms of your body's American youth
and wonder how Lesbia's precious pet
ever alit on that tiny blue bough so
near your eye's own beauty and wit.

I only know that summer sang in me
A little while, that in me sings no more.

Edna St. Vincent Millay

NYMPHS

In an infinite blindness of promise
they feed on the roots of aged trees
that imagine they are sequoia giants
until age seventeen when they emerge
full grown, as if thrown from a god's
tense thigh column, so many of them

over the centuries of earth, I assume
twilit clouds winging from the corpses
of my scavenger dive into snake skins
watching her sublime Oriental features
slip into a slight smile at her dinner
(secretive habit or shyness), profile

inclined by custom below the fiery fall
gaze of a fond ghost, recalling a Bengali
girl (cooler in her trendy "do" and Klein
jeans) a day or decade earlier, who took
pride in "giving good head," and a Moslem
bride quite able to bed the boy selected

by her rich family without shedding her
rough tongue rule of my gnarled trunk's
every trench and grey mesa, but she can
only eat so much (mouth hidden by a fan
of fingers), leaving several baby shrimps
marooned in their own bloody afterbirths

so many of them, escaping the night soil
of cultures that bound them in imago silk
before they unveil breast arias and labial
moans for a host king who shares no past
with them but want, their frenzied litanies
tearing at my heart as I carry them aloft.

Fruit cannot drop
Through this thick air.

H. D.

DREAM IN THE AFTERNOON

A small ball of auburn dog
on a shelf that does not exist
in the bedroom no longer mine
too near the ocean's tolling din
like a porcupine pup, but fluffy
luminous as an illustration
recollected from childhood

I gently cradle and carry
into an unrestored farm kitchen
where furious women and girls
materialize out of air and mind:

Amy alive among them, doll in
hand and desperate to be saved
Aunt Olga glaring in a teenager's
shiny red and green raincoat
the crowd of them multiplying
around us as they condemn me
without ceasing their chores

Amy's panicky hold on my arm
knocked loose by their bony
bodies' pinball batterings.

The dog could be that papillon
the old widow loved to parade
when I walked luncheonette
planks after school and football

or a fabulous bestiary notion
imported from medieval France

except for those round, brown,
disturbingly familiar eyes
and flick of a quiver as I lift her
free of the fierce female torrent

like the baby rabbit the cats
chased inside for me to rescue
and release in a nearby copse
the winter both my parents died

only noticing the dried blood
on my hands at the dinner table
after the enormous seashore sky
had swooped down with darkness
to sweep away the fir-tree towers
hugging this house to the earth.

RIME ON THE ANCIENT LECHER

Ghosts drying hard
wetness here, wetness there
across spines and bleached jeans
and pinkish daisy-dappled panties
in the outlandish office

(stained by dust and stolen smoke)

the unimaginable real scene
embracing her incredible flesh
curves from behind to kiss her
neck and unwinding tongue
one hand slithering down the fur
treasure cleft to gaff a liquid
clam shiver and sticky prize
the other free inside a silk bra
rolling a nipple hard as a coin
with thumb-evolving apery
as my penis achieved a size
only recollected in taverns
and harpooned itself between
the bounding worlds of her
well-conditioned buttocks.

Wetness, wetness everywhere

even as she stays virginal
in the coy space between

text and context.

COSMIC COSMETICS

Nature abhors a vacuum, skittering
away from light on centipede legs

and the universe's fly-wing flesh
is expanding into a graveyard
of black holes, I explain

to the beautiful girl a few years
past puberty who stars the half
moons over her eyes and dyes
her hair the color of fire

the mega clusters of restrung
galaxies may be a chess match

or meta arcs of the perfect circle
every hermit artist struggles
to marry in airy solidity.

I mouth the clipped flames of her
and fondle her pear bottom self
against mine in despair that
disorder's order is not

the paints she wields to stroke
a mad mother's dilating mirror
a father's pasteboard masks.

Go mad I cannot: I maintain
The perilous outpost of the sane.

Herman Melville

T. S. ELIOT DOES ZAHEDA'S *KAMA SUTRA*

The positions of love are extreme only in
their labors to regenerate or eliminate
the star shapes separating heads and limbs
where tongues sacrifice speech to a hunger
for escape from terse tombstone equations
and a pistol sun's geometric progressions
by licking hairy halos free of cosmic dust.

The women come and come again
around their aerobic totems.

As I shower behind sly Venetian winkings
each morning, a blind girl across the street
rocks a statue of herself from side to side
astride a frivolously red and white staff
that pierces me with voyeur sin and relief
her visage eroded into Medusa's stone rage
as I try to shut off imagining a Stygian
waste where word and image have no womb.

A wrinkle in space, in time, is the odd way
a brain yeasts when clenching itself into *me*
with a specific history but no thing weight—
and simplicity and Grecian symmetry succumb
to the clubfoot manifestations of some thing
that infects nature's every lurching nuclear
form with a compulsion to punch the mirror
cell impelling me to face alone the random
quanta of old cheese flesh, entropic sags.

Sitting behind Bickford's mass art
Cornell munches passing Tina tarts.

The Classics Professor chaired by iron legs
sported a cartoonish bowtie and boyish wit
in the college on the hill where salvation
was a secular noon thrill of funny stories
from the Depression that left no medieval
gold leaf of pious slime on his clumsy crab
departure from a room and a clogged earth
doomed to forget his life's gallant dance.

Pleasure *sans* guilt is not pleasure at all
only a muscle reflex that swats away horse
flies in a bookroom's desert places, but
this daily aging into the aches and abrupt
angst shudders of a body without memory
or weather nudges me nearer a slim Chinese
girl elevated by platform shoes and black
onyx slacks into an adult's chic flirting
where I nuzzle hair falls of empress silk.

Pieces of Prodicus to shore up
the glass ruins of sand floors.

After school, the Muse is a former nurse
and nun sanctifying the bed next to Aunt Ida's
who rises naked to cane herself at a trot
to the toilet she never reaches in time
her spine a storm-tormented fossil tree
sibilant as an Aeolian harp's cobra sway
even as flabby breasts flap against bone
regrets over never having been a cradle
to child or man, except in the dream she
wrings from the rosary of a purging god.

I must kiss them both, their fuzzy nods
and that of a wild Russian mother-in-law
stung mild by drugs and Long Island WASPs
above Byzantine bed sheet drape of folds
and ceases painters copy as a challenge
equal to the avalanche of buttery light
bathing a Mandarin girl's cheeky smile
before escaping the wind off Flushing Bay
for dinner with a small Bengali beauty
whose pouty and full sensuous lips always
laugh at my jibes and jigsaw reflections.

Kali's fangs and girdle of snakes
cannot conceal a moist cunt's quakes.

Meaty details of a recent drunken orgy
she shares over wine pointillates a wish
fantasy that will pillow sleepless nights
the three of them in a single bed (shot
from atop the grimy room Welles furnished
for his bravura Mexican strangulation)
the man entering her from behind as she
French kisses a Spanish girlfriend below
her, then sliding down to lap through fur
and eye-filmy walls toward the temple gem
that requires a dildo from a handy drawer
to force open whatever door remains between
them—not the man shrugged aside by their
camel contortions and playground squeals.

After such knowledge, a cat calls out night
and I run to the courtyard window, hoping
not for a father's ghost or Schroder's
Cheshire fading, but a stray stalker of food
another plump Mrs. Podge or Godfrey sidekick
whom I can tell about my fast rotting teeth

(now as false as the Feast of the Epiphany)
or teach the vulgate truth that the Romans
if never as brilliant as the unique Greeks
were more human in their readiness to risk
looking-glass speculums, to mourn with ash
tears a brother fallen on a foreign shore
and lay with women more than their match.

In her cave is my beginning
the beginning of my end singing.

Jewels in my mouth will not ignite serpents
or warn a deserted wife that global warming
can be an ego's dying upheavals, leaving her
adrift on an arctic chunk of art's amoral
immortality, or the anal white-out passion
of an American-spawned Oriental adolescent
but pearls inside oblique eyes will lantern
the ferry across Seng-Chao's illusory sea.

A poetics of death, not beauty, dribbles
from a dry brain in a dry season, love lost
in shedding what spring rain or reason divine
of a wife's lone winter rose, a blind girl's
keening the melody of a planet's obsessive
turning away from colliding rainbow nebulae
and the small yellow bus hauling her home
to fuel the lamp heart of a daughter dead
before death was born, unquiet as the fire
of a coruscating dragon when the elegant girl
from the East presses the full gift of herself
against the seven skins of my animal shroud.

The children in the tree go mad.
O, that prelapsarian rag!

OBJET D'ART

For Ernie Pappas

Blackness round as telescoped night
is neither black nor round
as a barber's bowl of moonless
arctic sky stained by stale polar
bear blood, cupping a sleepless head
that drinks at the classic breast
of a beautiful girl embarrassed
by its scooped cream nakedness
into a monosyllabic sulk.

The skull does not dream, my dear,
does not rave or rake leaves off
Nana's grave, fart in the face
of a geezer's ferret visage
chiseled from the marbled jowl
sag of a boyhood friend—elevated
to "Ernestos" by the bearded priest
who slips and falls in muddy slush.

A Grecian urn cradles love's ashes.

94

FINAL FUCK

Fantasize this: clenched
around the Mexican or
Indian girl's whole mud-
slick body under spilt Alamo
sheets of flat blood runs
and masculine rubble.

It become impossible
to isolate the specific
set of urgently undulating
mounds, to separate self
from female swell, foam
pleasure rising into tidal
fists that flatten
all geographies.

When a grey horror film
replays a sneak squeeze
and arched cat release

the insane doctor cackles
at a lightning-slashed
balcony, cod tongue deep

inside a welling mouth
mustang penis spurred up
the very tree of her

as a pilgrim finger trudges
the oily road to Damascus
hoping against reason

that the three will collide
in a sizzling spectral nova
that atomizes grave fears.

On the other hand, there
is a much funnier scenario
written for contracting
private parts that has

a don fork a mule
down infinity's Milky Way.

The sea slides back.
The mirrors are sheeted.

Sylvia Plath

THE HEART SHUTS

A hand mirror a week
from a five-and-dime store
(preferably black
to curb profane talk)
first at the home
then the hospital
overseeing the old
World's Fair lake

dropped or slammed
in a fury that would
abruptly ripple Olga's
filed patrician features
into Tragedy's trite
and toothless anguish.

But at last we relent
into gentler obituaries
of people and gossip
(giggles a shared itch)
remembered from Sunday
feasts around Nana's table
rafted here from Italy

before CIA operatives
began listening at vents
for Aunt Emily's glass
slippers to shatter
at our feet.

Three silences there are: the first of speech
The second of desire, the third of thought.

Henry Wadsworth Longfellow

EVERY THIRD THOUGHT

On a hill above highway-wrapped
Queens and the honeycomb heart
of the family plot Grandpa bought
in 1929 to bury his first-born
son, dead at age 22 and bearing
my name into the New World and
into the sepia photo Nana kissed
in place of the Sacred Heart's
3-D torch each night and morning
we encircle the reordered earth
with plastic roses in our hands
to strew atop the hermetically
sealed steel box, refrigerating
Emily's flower-fragile remains
before it rockets to the sphere
she loved with unrequited lust.

A lady of many Romance tongues
(girlish in a painted grimace
and shocking sunset locks
Olga used to tend and tint)
labors to architect a statement
of momentous steeple intention
but can only nail down "pretty"
with a child's obdurate force,
and no degree of robot coaching
can coax synapse into a syntax
until—millenniums later—she
catapults "sister" into my face

and I am flung back from the sea
where she drowned with an infant
who died before my conception
from the disease-tainted touch
of a drunken doctor's house call.

The ride home is the ride home
through familiar neighborhoods
now occupied by foreign troops
of mothers and toddlers brazen
in their assumption of our past
as if we had not existed there
in the apartment flats and narrow
houses wincing under the hatchet
blows of their exotic languages—

only the woman in the night car
able to unloosen knots of loss
with her wordless body and lips
welcoming mine as rooted thighs
rock us here, and her kind smile
moons over panicky diggings.

THIRTEEN WAYS OF LOOKING
AT AUNT IDA

1

Alert eyes and benign smile
follow me with the alarm
of an animal at bay
Ida's hand bent into a paw
the other trembling
metal bed bars.

2

Ten billion neurons misfire novas
when Ida writhes to launch a word.

3

They promoted Ida so far ahead
of her class, Grandpa had fled
his busy barber shop one June
to beg the glacial Irish
principal to let her
remain his little
girl a little
longer.

4

A palimpsest portrait of an Aunt Ida.

5

The Roman lady in a flowing pink toga
painted from behind on a field of green
pinching flowers from a starry stalk—
Ida loved beyond art's abstract frame
because it could be seen from the table
every meal after Nana died and Grandpa's
blood had embarrassed his new twin bed.

6

Ida's legs, thin as lath under Alp
sheets, pierce the Jello cosmos
of laws that Spinoza divined
leaking precious calcium.

7

Social worker and compulsive cleaner
Ida was the mother of none and too
many, of parents and mad sisters
and their fatherless broods
swept into Nana's house
and kept noisy and safe
from winter scenes.

8

IDA is the software of all consciousness
but I cannot AUTO SAVE a single atom of her smile.

9

Icon and ritual conflate in
Ida's wall-less last church
"Olga, Olga, Olga," her sole
chant and mass for the dead.

10

Books were the bricks of Ida's tower
and exact instructions for future eyes
about foods and funerals and family
charts pinned to jars, shirts, unopened
gifts, as if closet and shelf lives
would never yield to roach babble
the abyss of a silent forever.

11
Form follows function into Aunt Ida.

12
Cribbed in Aristotle's arms
Ida has been reasoned down
by history or obscene whim
into a pet that can shit
but cannot feed itself.

13
The so slow crow cold turning
away from the light and me.

Dark Mother, always gliding near, with soft feet,
Have none chanted for thee a chant of fullest welcome?

Walt Whitman

HALF MAST

The dreams continue into summer.

Disappearing borders trench
the swamp that men age into
when a century's spine snaps.

What face clamps my cock so hard
it quenches the fever that never
achieves a barracuda leap?

The moon's almond after-image dangles
from ceiling dust with an Oriental
stare, gentle but feline-aware, a web
of infinite lines threatening
hysteria, unbearable joy:

I am earth, universe, orchid cunt.

black and white prints, near sepia
from the American 40's when June
Allyson and Doris Day palmed automated
kitchens that ejaculated lace-curtain
brides and a doll's spotless apron
to wipe the cream cum off floral
bedroom walls, when Kate Hepburn
could blueball you for hours
while pruning Joyce's trellis
and planning her next abortion

rally, Bacall going down on Bogie
for the third time, although
ambivalent Bette Davis crouched
behind a couch in broad daylight
like an aging ingénue, knives
of delight in her white-gloved smile.

Who lies spent beneath me
when a tattered dog star slinks home?

I am earth, sister, membrane veil
you slash night after night.

I remember my father saying a year
before his death, "Outside I look
different but inside I am me.
It's crazy as that kid who eats
dirt and plays with himself
no matter how many times
his old lady belts him."

Plum nipples, butter mounds, wine
lips, a tidal drift of insistent angel
hair, stroking the flat edges of me
fat over the wrecked skull where eels
wave me in, unyielding while yielding
voices from outside blaring fog-horn
warnings of fatal tick feasts.

Hold still. Lie still. Be still.

Marmoreal will ascends in slow motion
under sacks of uncertain salvage
again and again and again
to glut a flame's liquid throat
to weld whole whatever relics

are tearing a rag body stiff
as a girl becomes a man becomes
a spilt infant becomes a raven flock
of women tonguing every wheat mile
of me to climactic heights where
I almost sweat free of penny arcade
pincers and the schoolyard pebbles
that clawed knees into laughs
scaling ice-stocking thighs

becomes a senile grandmother too
carelessly dropped into an armchair
beside the flickering subway car
screen to bark broken rosaries
of caramel teeth at the pair
of plump squirrels chasing

one another up the sun's
incinerating ribs.

I am earth, daughter, chaos come
round to wetnurse your dead.

Saint Anselm of the good ship *Nekyia*
writes on a log to the Monk of Fools
that that is how, fungus-wise, life
and literature always entwine
like fairy thorns of eternal
foreplay, three limbs aloft
without a cactus hope
of consummation.

I remember the last molar left
in my father's restless mouth
afraid of dentists, doctors
arctic kindness of any kind

a stump of dumb defiance
he worked loose with conversing
metal heads over several days
as thick pea soup (brined to tears)
perked like lava on a hot plate
in that soiled, narrow room
where bachelors and widowers
and non-union cowboys hunch to die
refusing to wear shirts or take
the yellowed cellophane shroud
off the single double-breasted suit
waiting in a closet confessional
like a punished delinquent
for a wedding or funeral march
or an unbelievable long shot
in the seventh at Aqueduct

Let me lick you clean of my filth.
Let me lap shut the next crack of song
that cradles only crocodile meat.

afraid of appointments,
computers, soap opera complications
but not of a slurred, a steady
a deliciously self-inflicted pain.

I am cloud to your brick shouts
moss to your stone keenings.

I'LL SUCK THE JUNGLE FROM YOUR SKIN
BY ITS HAITIAN ROOTS! I'LL SHIT YOUR
SEAS FOULER THAN AUSCHWITZ MUCK!
I'LL FUCK THE AIR FROM YOUR HIGHEST
LUNGS UNTIL YOU MEMORIALIZE MY NAME!

I am the gelid floor of your nothing.

Come swallow me now, you castrating bitch!

Cat face dawns at the window, meowing
forgiveness, a half-eaten mole
sacrificed at the porch altar where
the old evil walks, stirring me
erect. I let her in and recall
the ghosts of dolphin breasts
nudging him safe and awake, weary
as the Dali poster of translucent
sexless ladies haunting a parlor's
desert air before sinking back
into this heaving half-life
between mirror and shore.

CASTOR OIL

Sunday is every day now
vast and solemn
          as a vacant cathedral
                  riddled
by the stares of rainbow martyrs
                    until time staggers
loses its urgent silver tongue.

Being old brings the relief
                of failure and safe retreat
a slow burrowing
          back into eyeless roots
                  infant sleep
below alarming sirens
            factory hymns
                shameful hospital
sheets.

I listen to the Puerto Rican janitor
swab the vestibule floor
              (ersatz marble chessboard
                for absent aristocrats)
sea sloshing
       across a bloody deck
       as its sunk steel hull
             buckles under his heel.

I linger over elegant shadows
             of dinosaurs
             and saber-tooth tigers
drifting
      out the window to die
             tooth and claw climbing
ladders of smoke.

But the stench of the room is armpit ripe
                              with disaster
a single survivor floating
                    in the wake
                              of a behemoth god

    whose huge sour heart I can still taste.

NO MAN'S ISLAND

Smash the stone image.
Gut the old man's bark.
I'll go no more aroving
down to a mercurial sea.

NOTES

"Preface" September 30, 1999 is the author's
birthday and termination date for these poems—hence a
last collection. If true, it suggests a grandiose effort to
impose a single mortal ego upon history; if false, it
represents a no less egocentric attempt to plant Wallace
Stevens' barrowed jar in a roiling ocean floor. In any
case, author and audience are fated to melt into the vast
folds of eternity—a stentorian noun so terrifying it
really cannot be heard—as implacably as rain washing
diary pages blank. Embedded with threes and pleasing
to pen, magical as they come, 1999 is thus at the ax
edge of momentousness and the author's ebbing
existence. It was a final year in which seventy percent
of his country's population growth consisted of
immigrants, about a million in all, with Flushing,
Queens—his once and always home town—boasting a
population that was already half Asian. Near year's
end, months after the book's completion, two stories on
the front page of the *New York Times* (December 23)
provided a jagged pair of relevant parallel texts.

In one, scientists in Oregon report insinuating
jellyfish genes (presumably without their permission)
into monkey embryos, which is viewed as a positive
prelude to stuffing genes into human embryos. Still on
the West Coast, which can act as a sewer basin for the
nation's cloudier run-offs, the other story (from San
Francisco) concerns five homeless men being beaten to
death in Denver and two others having their heads cut
off. A homeless man in Richmond, Virginia was
beaten and stabbed before being decapitated, his head
carried away to festoon a footbridge.

In two other cases, the homeless men brought their
troubles upon them-selves. A vagrant in Dallas, Texas,
was blasted by a shotgun at close range for going

through someone's trash, and in Chico, California, a companion in need was punched to death for asking the wrong person for his spare change. Such attacks seem to be multiplying, the National Coalition for the Homeless claiming twenty-nine victims were murdered in eleven different cities over the year. The attackers were usually young men seeking to alleviate boredom and stifled rage by "bum-bashing" and "troll-busting," which might be a crude salute to poetry's residual power among the illiterate.

On Thursday, July 8th of an unusually hot summer, the police arrested two black men for beating a white homeless man to death outside Thompkin's Square Park on the Lower East Side of Manhattan. The victim was Joseph Radu, Jr., age forty-four, who had been hospitalized on June 13th after being mugged in Queens, a beer bottle smashed over his head. When he left the hospital a few weeks later and returned to the park, he was still wearing a hospital gown and slippers.

His attackers had been drinking in the park, which enforced a midnight curfew and was avoided by local residents because of its large homeless and rodent populations. They had argued with some skinheads earlier and were in a foul mood when they encountered Radu sleeping on the sidewalk next to a homeless woman, waking him with nudges. He cursed them, using racial slurs, and they began kicking him, then slammed his head against the concrete.

By the middle of 1999, it was believed that three hundred million people had perished from small pox thus far in the twentieth century, and the front page of the Saturday *Times* (August 28th) would estimate that malaria slaughters 2.7 million people annually, mostly children in underdeveloped countries. This same blood-soaked century had witnessed the destruction of over eight million combatants in World War I, followed by an epic influenza epidemic that killed twenty-five million more. Fifty-five million people, civilian and military, lost their lives during World War II, which climaxed with the explosion of two atom bombs.

In apparent contrast, a story on page A10 of the August 5, 1999 edition of the *Times* noted that the population of India, "which adds more people to the world each year than any other nation," would officially reach the one billion mark on August 15th (Indian Independence Day) and overtake China within four decades. By October 12, 1999, the global population had reached six billion human beings. The author learned from a television documentary that there are over two hundred million Wongs in the world: How does one assimilate that many Wongs?

Death is the surest heritage and crime of human birth, an abstract of an absolute abstraction —like infinity's continuum—incapable of being bound by language or tamed by philosophy, though there is a rational, hence sentient, consolation in draping it with suits of sound and umbrella meanings ("nothing," "sleep," "oblivion," *et al.*). Paul Celan (*né* Ancel), holocaust orphan and suicide in April of 1970, configured his poems as "messages in a bottle," which may also apply here.

The stomach-churning fear that death of self inspires from the outside intensifies into bedroom terror as the anticipated vacuum spreads nearer, its approach often preceded by the disappearance (*forever!*) of beloved others and the steady, painful breakdown of organs and joints. The mind itself, elusive ghost in the elaborate electrical machinery of the brain, orbits inside its cozy helmet of bone with restless creativity, rotating ever closer to an increasingly present *and* distant past, invented or real—or both and merely rearranged.

For the poet, the artist, the intrinsic atrocity of human existence, our ken of its inescapable erasure, is more horrific and much earlier experienced—detaching self from existence in the process—because of his or her blighted childhood (loss its signature *donnée*) and consequent emotional fixation. What makes the torture so severe in later years is the daily knowledge, seeping inside at unforeseen and more frequent moments, that all one has touched and tasted and loved, including Wordsworth's luminous childhood "spots of time," will be wiped out as totally as any gallant Hemingway quote—"grace under pressure"—once scrawled on a high school blackboard.

Quoted in Maurice Yacowar, *Loser Take All* (New York: Frederick Unger, 1979), p. 209, Woody Allen scanned his past to affirm, "Maturity has borne out my childhood. I'd always thought Death was the sole driving force: I mean that our effort to avoid it is the only thing which gives impetus to our existence." Also in 1979, Christopher Lasch's *The Culture of Narcissism* (Norton), p. 208, thought "the fear of death takes on new intensity in a society that has deprived itself of religion and shows little interest in posterity."

Tough-minded Arthur Schopenhauer (1788 -1860), son of a suicide, was uncharacteristically lush in his précis of our fated journey: "We begin in the madness of carnal desire and the transport of voluptuousness, we end in the dissolution of all our parts and the musty stench of corpses." This is from *Essays and Aphorisms*, edited and translated by R. J. Hollingdale (New York: Penguin, 1970), p. 54, who notes that the philosopher, "like many very intellectual men...had a strongly sensual side to his nature" (p.33) Refusing to see his mother after 1814, Schopenhauer was a misogynic user of women, who had the courage to voice the reality that our "existence has no foundation on which to rest except the transient present" (p. 52).

Since childhood, when he scissored history articles from a *New York Journal-American* series and accounts of Brooklyn Dodger games from the *New York Daily News* for his scrapbooks, the author has been fascinated by obituaries. They were extensions of history (a favorite subject), people and events that textured the past and his own days, rendered existence larger than self-reflection. In 1999, conceived as a plague year, he reverted to childhood's habit, clipping obituaries from the *Times*, along with material from other journals, as entries for his *Doomsday Book*.

Saturday, February 19th of 1999, Sarah Kane, a twenty-eight-year-old playwright, hanged herself in a London hospital, where she was being treated for severe depression. Kane, who lived alone, had seen her first play, *Blasted*, produced when she was only twenty-three, a violent drama in which a rampaging soldier rapes another man, gouges out his eyes, and forces him to cannibalize a dead baby.

The *Times* obituary of February 25th quotes Kane's unique take on her illness: "Many people feel that depression is about emptiness, but actually it's about being so full everything cancels itself out. You can't have faith without doubt, and what are you left with when you can't have love without hate?" That same Saturday in New York Hospital, Willard Espy died at age eighty-eight, an elegant, handsome man who had once been a model in Paris, an inveterate punster possessed of a whimsical smile whom the author met several times but did not really know.

Espy had worked at the *Reader's Digest* from 1941 to 1957, marrying and divorcing twice before moving to East Hampton in the 1960's. After his first book, *The Gift of Words* (1971), he produced nearly a book a year through the rest of his life. His most quoted lines are: "I love the girls who don't./ I love the girls who do;/ But best, the girls who say, 'I don't.../But maybe just for you.'"

On Monday, February 22nd, poet William M Bronk died at age eighty-one from emphysema in Hudson Falls, New York. Following a small-press career of little notice—he did not travel or give readings—his collected poems, *Life Supports*, won the American Book Award in 1982 in spite of its proneness to abstract language and thematic focus on metaphysical subjects. A graduate of Dartmouth and veteran of World War II, he had run the family business, a coal and lumber company, until he retired in 1978. He left no survivors but was a descendant of Jonah Bronk, whose farm gave Bronx County its name.

Page 3 "Intimations of Mortality" The epigraph is from Aiken's "An Old Man Weeping," which can be found in his *Collected Poems: 1916-1970* (Oxford University Press: New York, 1970), pp. 462-64.

Besides marking the end of World War II, 1945 was the year that the author's mother, Lena Butscher (*nee* Angelina Cutillo), age thirty-nine, was removed by ambulance from her home, a ground-floor, three-room apartment at 46-06 Smart Street, Flushing, as a small group of neighbors looked on, her puzzled middle son among them. Taken to Queens General Hospital, the Author's birthplace , she was diagnosed as paranoid schizophrenic and transferred to Central Islip State Hospital on Long Island, remaining there for virtually the rest of her life. Her sisters, Edith and Mabel, filed a petition, alleging that the author and his two brothers, John and Ronald, were "neglected children."

On August 1st, the author and his older brother John were "paroled" to Edith, Ronald, then only two years old, went to Mabel (Farrell), a war widow sharing a third floor apartment at 43-76 163rd Street with her

116

daughter Genevieve ("Ginger"), though later the author went to Mabel's, Ronald joining John at Aunt Edith's home, the first floor of a corner house owned by a Mrs. Horowitz at 33-23 145th Place in Flushing, which was also home to the author's maternal grandparents and Aunt Emily, "Olga" in the family (*née* Emma Olga), who, like Edith, worked in Manhattan.

Next to the Smart Street apartment building was a small wooden house with plate glass windows that functioned as a grocery store. Owned by Mrs. Rocca, a small, dark, squat widow tented in a standard European black dress, it had the saving grace of selling cigarettes to minors, which facilitated the author's precocious introduction to the habit—his father smoked his Camels down to his burnt finger tips with a prisoner's cupping hold, and Aunt Olga chain-smoked her Parliaments, further filtered by a very feminine cigarette-holder, smartly to her breathless end.

In his fifties, the author's father would confess that he could no longer "get it up," but blamed it on a stalwart Catholic loyalty to a wife who would not let him visit her "in the nuthouse," explaining, "You gotta use it or lose it." But it was at Mrs. Rocca's, ushered there perhaps by his silent mother, that he saw his first dead body, not at the funeral centering "First Death," which came years after and which he did not attend.

Mrs. Rocca's chubby, black-haired, rouge-cheeked daughter or niece, dressed and lace-pillowed in communion white, framed by urns of flowers and two tall candles, was on display in the parlor behind the store. In an unpublished novel, the author had tried to re-imagine his feelings, or lack of same, as he knelt and made the expected prayer before the small ornate casket and could not stop thinking about the shelved rainbows of candy and gum pulsating in the front room.

Much later, it was just such isolated childhood incidents, suspended like animated dioramas in the amber of the author's memory, that confirmed how repression can protect the psyche against disturbing memories too alive for safe burial by severing their emotional chords.

5 "First Book" The book was a hard-cover volume of Uncle Wiggily stories given the author as a birthday present in 1945 by his brother John when they were still living together in the Smart Street flat, their father having hired a strange little woman named "Etta" to clean, cook, and be there for them—possibly the result of a court order. The second book remembered by the author was a gift from St. Mary's Elementary School, a few blocks away and just around the corner from the wooden church where he lost his faith, as reconstructed in "First Sonnet." Near year's end, the nun teacher told the author and his classmates they could come up one at a time and take a book to keep from the array of library discards arranged along the radiator top. He chose a thick, glossy volume about King Arthur and Knights of the Round Table, illustrated by N. C. Wyeth.

Books were always covert, joyful ways of not being with the world that healed a wounded self into Byronic demigod and Joycean nay-sayer. With innumerable other scribblers and gipsy scholars, the author joined Jorge Luis Borges in "The Library of Babel," a universe where Borges, growing blind, could write, "Perhaps my old age and fearfulness deceive me, but I suspect the human species—the unique species—is about to be extinguished, but the Library will endure, illuminated, solitary, infinite, perfectly motionless, equipped with precious volumes, useless, incorruptible, secret."

This might represent whistling in the dark of an empty church, but it can soothe an unquiet mind with the narcotic of language distilled from legions of now unremembered books. See Jorge Luis Borges, *Labyrinths: Selected Stories and Other Writings*, edited by Donald A. Yates and James E. Irby (New Directions: New York, 1964), pp. 51-58.

7 "Barring Decapitations or Kind Embalmings of Sudden Acclaim" Raised in Britain, actor George Sanders (1906-1972) came to Hollywood in 1937 and established himself as the Saint and a stylish villain with an upper-class accent and manner via a string of entertaining, if minor, films, although he did star in *The Moon and Sixpence* (1942) and had a major role in *All About Eve* (1950).

"Macdonald" is Ross Macdonald, pen name for the mystery writer Kenneth Millar, whose daughter Linda was cause of much grief for him and his wife Margaret (*née* Sturm), also a mystery writer, whom he had wed in in 1938. Many of Macdonald's novels have Freudian themes, as in *The Galton Case* (1959), and he himself underwent intense psycho- analysis for two years (1956-58). Millar was diagnosed with Alzheimer's in 1981 and died at age sixty-seven on July 11, 1983.

The William Faulkner episode, conflating a fall from a horse named Stonewall and a subsequent social meeting with a doctor friend as a single event, is from Joseph Blotner, *Faulkner: A Biography* (New York: Random House, 1974), II, p. 1829. Faulkner, age sixty-four, died not long after from a "coronary occlusion" at 1:30 a.m. on July 6, 1962.

8 Mother Teresa of Calcutta (*née* Agnes Gonxha Bojaxhui) was born in Yugoslavia but earned her reputation as a missionary working in the slums of Calcutta, opening her House for Dying there in 1952 and establishing a leper colony in West Bengal five years later. She was awarded the Nobel Peace Prize in

119

1979 and died in 1997 at the age of eighty-seven. For a discerning critical view of her career, see Christopher Hitchens' *The Missionary Position: The Ideology of Mother Teresa* (1995). Certainly, she could have given far more to India by defying her church's position on birth control than relieving the miseries of a limited number of the sick and poor with attendant, self-aggrandizing publicity.

Russian-born actor Yul Brenner (*né* Yul Bryner) died in 1985 at the age of sixty-five from lung cancer and in the last years of his life was a dramatic spokesman for the American Cancer Society against smoking.

9 The allusion to Lucifer is patently to the Lucifer Dante depicts in the final canto of *The Inferno,* a gigantic three-headed monster trapped in an ice field, chewing on Judas, Brutus, and Cassius and beating his huge wings futilely against the polluted ice produced by rivers of guilt. See John Ciardi's fluently vernacular translation, *The Inferno* (New American Library: New York, 1954), pp. 282-87, a sturdy paperback treasured by the author since high school. He met Ciardi one summer at the Breadloaf Writers' Conference and was impressed by his absorption in the stock market.

12 "Painter Man" Chicago-born (October 29, 1908) John Opper was part of the Abstract Expressionist movement that shifted the center of Western art from Paris to New York after World War II. His Cubist-inspired pictorial painting style eventually evolved into pure abstraction as he became convinced while living and teaching in Wyoming between 1947 and 1957 that no artist could match nature, thereafter concentrating exclusively on color and a few simple forms. Abstract artist Agnes Martin claimed, "It took me twenty years to get beyond nature," and Odilon Redon, associated with the late nineteenth century's Symbolism and

Decadence movements, said colors "contain a joy which relaxes me; besides, they sway me toward something different and new."

In his seminal *Modern and Modernism,* Frederick Karl, friend of the author's, observed that "when the unconscious entered the vocabulary of every major artist, it became apparent that the inevitable goal for Modernism was, in one sense, stream of consciousness, as it was, in another sense, abstractionism."

John's attractive wife Estelle (*née* Hausman) devoted herself to their children and his career, at some cost to her own ego needs as they moved from college to college. After Manhattan, where he first migrated in 1933 to work for the WPA, studying with Hans Hoffman, and returned to in 1957 to teach at NYU and become part of the art scene revolving around the Cedar Tavern, the Oppers' last home was Amagansett. Estelle died there in July of 1994, and John followed several months later. They are buried in the Green River Cemetery, East Hampton Springs.

16 "When the Living Is Easy" The Bryant epigraph is from his well-known "Thanotopsis," which struggles heroically against enormous death pressures and, at poem's end, converts terror into a Whitmanesque embrace, death morphing into "pleasant dreams."

Context is all and nothing, the seeding of the earth with human remains as remorseless as moss thrown on churchyard tombstones. Monday, March 29th, Vera Tolstoy, granddaughter of the Russian novelist and the last living link to him, died at a retirement home in New Smyia Beach, Florida, age ninety-six. She was seven in 1910 when her grandfather died and spent her teenage years in Czechoslovakia, but her careers included hair-dressing in the Prague, singing Gypsy songs in Paris, and selling perfume in New York, having migrated to the United States in 1949.

April 11th in Belgrade, Serbia, newspaper publisher Slavko Curuvija, a tall, suave man with a small grey beard, and his wife, the historian Branka Prpa, were returning to their apartment after an Easter lunch out, when two men dressed in black shot him in the back and pistol-whipped her. Before leaving, they fired several more bullets into his head. In recent weeks, the Yugoslav government had moved to shut down all independent media, and Curuvija's paper and weekly journal had been heavily fined the previous year for breaching restrictive information laws.

17 "Italian Pops" Giuseppe Verdi (1813-1901), whose first wife and their two children died between 1838 and 1840, married former operatic soprano Guiseppina Strepponi in 1859. Along the way to fame, he produced several illegitimate children, who were deposited in local orphanages.

20 "Ignatow Eats the Sunset" David Ignatow (*né* Ignatawsky) died at his East Hampton home on November 17, 1997, age eighty-three. His wife, the artist Rose Graubart, whom he had married in 1940 when living on the Lower East Side, preceded him, dying of cancer on May 12, 1995, a Friday, and is buried in Cedar Lawn Cemetery in the heart of East Hampton, alongside her son David, a brilliant student institutionalized with schizophrenia in his twenties who died in 1985.

Ignatow had expressed the wish to the author and others to be buried in Queens with his family—"I would be buried beside my parents," he wrote in *Shadowing the Ground* (Hanover: Wesleyan, 1991), p.

38—across the street from the high school where the author taught English for thirty-one years. But the present location of his ashes, presumably in his daughter Yaedi's possession, is unknown.

The only son of Jewish immigrants from Russia, David Ignatow was forced by the Depression to quit college in his freshman year, though his father, a printer with whom he fought frequently, did publish his first collection, *Poems* (1948), which was warmly reviewed by William Carlos Williams in the *Times* (November 21) and helped launch his teaching career. Ignatow had a mistress in his last years but continued to share a house in the East Hampton Springs with his wife and daughter, although his half was run separately from theirs. Graubart, who had a surreal or at least a Chagallian sense of humor, dubbed him "the pointy-headed fuck" in conversation and letters to the author's wife, Paula Trachtman.

One sunny weekday, after disembarking from the Hampton Jitney in front of the Huntting Inn, the author ran into Ignatow in the middle of his doctor-prescribed morning constitutional up and down each side of East Hampton's Main Street (Montauk Highway) and told him he was thinking about writing an essay on the underlying psychological similarities between male poets and serial killers. Ignatow laughed and claimed he had recently begun a poem in which he wrote, "I wanted to write a poem. I got a gun."

21 "'I Do the Police'" See the facsimile edition of the original drafts of *The Waste Land* with Pound's annotations edited by Valerie Eliot (New York: Harcourt Brace, 1971). The title was borrowed by Eliot from Charles Dickens' *Our Mutual Friend*: "He Do the Police in Different Voices." The artist in the poem may be Robert Gwathmey, who was wed to Rosalie, fellow

Virginian, and who specialized in flat, bright, neo-primitive paintings of native Southern scenes with a reformist edge—he was often assumed to be black. Although the incident described is probably fictional, the artist, a congenial bourbon drinker and droll story teller, suffered from Alzheimer's in his later years, spending the last months of his life in the Southampton Nursing Home, where he had to be tied to the bed —he died September 20, 1988, age eighty-five, Rosalie living on until December 26, 2001.

The author and his wife met them a decade after they moved to the Amagansett house on Bluff Road built for them by their architect son Charles in 1966, a showcase for his softer-curved version of Wright modernism.

34 "Hog King of America"   The title and content of the poem are taken from a feature article the author read in the *Times*, since discarded or lost.

36 "First Date"   When the author was growing up, there were three movie theatres in downtown Main Street, Flushing: the RKO Keith's, where Main flows into Northern Boulevard, a genuine movie palace that ran old movies every Tuesday and Wednesday; the Town Theatre a few blocks up or south, known as "The Itch," specializing in old films and serials, along with the regular two new films every week; and the Loews' Prospect still further up where Parson's Boulevard peeled off from Main, across from the triangular site of the library given to the town by Andrew Carnegie, a solid neo-classical structure since replaced by a much larger, block-like modern building. The Town was indeed demolished, though the author never saw the wreckage, and the Prospect converted into several stores. The Keith's (baroque interior destroyed illegally by its Korean owner) is walled shut and unused. It was the Prospect where the author took his shapely cousin, who had a crush on him, and the movie they watched (or didn't watch) was Cecil B. DeMille's *The Greatest Show on Earth* (1954).

In 1999, a number of cinema figures faded to black. Among them was Garson Kanin, who died on March 13th, a Saturday, after a lengthy illness at age eighty-eight. He had been married to actress and screen-writer Ruth Gordon for forty-three years when she died in 1985. Together, they had written *Adam's Rib* (1949) and *Pat and Mike* (1952) as vehicles for Katherine Hepburn and Spenser Tracy, and he had been the author and director of *Born Yesterday* on Broadway in 1946, starring Judy Holiday, which ran for three years before becoming a movie. Kanin is quoted as admitting, "I become physically ill if I don't work for three days."

A young actor named David Strickland, who was a regular on *Suddenly Susan*, a television sit-com entering its fourth season, died on Monday, March 22nd, an apparent suicide, found hanging from a bed sheet in his Los Vegas motel room. A native of Glen Cove, Long Island, he had been arrested in October the year before for cocaine possession and recently stopped taking the lithium prescribed to control his mood swings.

May 25th, Hillary Brooks, a classy character actress born in Queens and a graduate of Columbia University, died in California at age eighty-four. She had been in a variety of films, ranging from *Jane Eyre* (1949) to Alfred Hitchcock's *The Man Who Knew Too Much* (1956), although the author associated her with the TV series *My Little Margie*, which ran from 1952 to 1955 and which he would never admit to watching.

At about 10 a.m. on a quiet Sunday morning, June 20th, a bull named Roughrider escaped from an illegal rodeo and charged through the streets of Long Island

City. Four policemen pursued him in their car, shooting forty bullets into him before he collapsed.

Sylvia Sidney (*née* Sophia Kossow) died on July 1st at Lenox Hill Hospital in Manhattan, age eighty-eight, after an acting career that spanned from the silent movies of the 1920's to the present and included starring in Fritz Lang's first three American films, *Fury* (1936), *You Only Live Once* 1937), and *You and Me* (1938). Born in New York, she was nine when her father, a clothing salesman, and her mother divorced. She attended the Theatre Guild school but was expelled for staying out late. Sidney's three marriages—to publisher Bennett Cerf, Yiddish actor Luther Adler, and Carlton Alsop, a publicity agent—ended in divorce, but she had a son with Adler, Jacob, who died in 1987 from Lou Gehrig's disease (amyotrophic lateral sclerosis). Examining her long career, Sidney explained, "I wouldn't know what to do with myself if I retired. I'm an actress, and I have to work."

On Wednesday, August 6th, Victor Mature died in California from cancer at the age of eighty six, if his Kentucky birth date of 1913 is accurate. A big, fleshy, hooded-eyed and handsome leading man who made his mark playing soulful Nick Bianco to Richard Widmark's vicious Tommy Udo in *Kiss of Death* (1947), Mature eventually appeared in over fifty-five movies in his career—*Samson and Delilah* (1949) and *The Robe*(1953) among his muscular Biblical epics— which he ended when he reached the age of fifty-six. Charmingly self-deprecating and a survivor of five marriages, he took pride and much pleasure after retirement in playing golf six days a week.

Actor George C. Scott also died on a Wednesday, September 22nd, in his California office at the age of seventy-one from a ruptured abdominal aortic aneurysm. Scott was the grandson of a coal miner and worked on an assembly line in Detroit during the Depression—his mother, a semi-invalid amateur poet who gave recitations, died when he was eight. He served in the Marines and was a heavy drinker but would give compelling performances on stage and in films such as *Dr. Strangelove* (1964) and *Patton* (1970) though proudest of his 1957 performance in *Richard III* for Joseph Papp's Shakespeare Festival. Married at one time to sultry Ava Gardner, Scott was accused of beating her on two separate occasions. Anent his career he confessed, "It's never been difficult to subjugate myself to a part because I didn't like myself too well. Acting was, in every sense, my means of survival."

38 "Apollo I" A fire inside the command module of the Apollo I spaceship on January 27, 1967 quickly killed its three astronauts, Virgil I. Grissom, Edward H. White II, and Roger B. Chaffee, on the launch pad, their death certificates specifying "asphyxiation due to smoke inhalation due to the fire" as a logical sequence.

43 "Another Passage to India." The epigraph is from Michel De Montaigne, *The Complete Essays,* translated M. A. Screech (New York: Penguin, 1991), p. 18. Montaigne also observed that "according to Aristotle, of all artists the one who is most in love with his handiwork is the poet" (p. 451).

The two cats were mother and son, one-eyed Mrs. Podge and vacant-eyed Pushkin, who was mentally damaged when he fell or was accidentally shoved off the daybed while his mother was giving birth to him and his three siblings, one of whom could not be expelled. The author departed from his apartment before the messy process was completed to attend classes at NYU in vain pursuit of doctoral knowledge, leaving his wife and her friend Sandi Newman to assist

Mrs. Podge and then drive her to a veterinarian on Queens Boulevard, who extracted the stillborn kitten.

Much removed from Zola's bright creature, "Nana" was the name used by the author and her other grandchildren for Jennie Vincenza Cutillo (*née* Alvino), who had emigrated to America in 1901 with her husband Charles from Bella, Potenza, a small hill town in the outskirts of Naples, where she had been born on November 8, 1878. Charles was almost twelve years older than his bride, who would curse him as "the old man" in her senile last years. The Atlantic crossing had been so sickeningly rough that Nana vowed never to repeat it and would bear ten children in the New World, two of whom were lost in childbirth, she herself dying June 19, 1955 in Flushing, a town she hated with an explosive fury because, unlike Manhattan and Corona, it lacked noise, ethnic neighbors—and because she was drowning in dementia.

45 "Sweet William's Conceit" The Nabokov quote is from his essay "Perfect Past," which the author took from *The Best American Essays of the Century*, ed. Joyce Carol Oates (New York: Houghton Mifflin, 2000), p. 303. An enlarged and revised version forms the first chapter of Nabokov's memoir, *Speak Memory* (1966). The entire sentence commences with an enviable image: "The cradle rocks above an abyss."

Emerson's metaphor is from his essay "The Poet," which is in the convenient Modern Library edition of *The Selected Writings of Ralph Waldo Emerson* (New York: Random House, 1950) assembled by Brooks Atkinson, pp. 319-41. As he also wrote, "Every word was once a poem."

In *The Origins of Knowledge and Imagination* (New Haven: Yale, 1978), p. 44, Jacob Bronowski lectures, "The idea that we can conjure the world with names, with nouns, even with verbs, is a familiar belief among primitive peoples." Polish-born Bronowski (1908-1974), a mathematician, government scientist, and popular lecturer for the BBC, located his Alamo metaphysic in the belief that "there are no events anywhere which are not tied to every other event in the universe," but was forced to admit we humans "*cannot* extricate ourselves from our own finiteness" (p. 70).

Currently popular among leading particle physicists, the string theory maintains that the cosmos consists of trillions upon trillions of vibrating strings or strands of energy, self-defined by the modes of their vibrations. If true, there can be untold numbers of parallel universes universes (megaverses), none of which can prevent Shakespeare's King Lear, who is fictional, from losing all that he loves and once ruled. See Stephen W. Hawking, *A Brief History of Time* (New York: Bantam, 1988), pp. 159-62, re heretic superstrings.

On Friday, May 14, 1999, Kyriocos Demetriou, a barber known as "Mr. Kay," died at age eighty while watching television in his East 86th Street home. He had taken over the Broadway Barbershop in the early 1950's, the oldest in the city, dating back to 1904 and located between 103rd and 104th Streets. Born in Cypress and raised in London, where he opened his first shop, Mr. Kay was famed for his straight-razor shaves and old-fashioned ways, playing classical music on the radio and giving his young customers lollipops.

The author's maternal grandfather, Charles Nicola Cutillo, a quiet, dignified figure with a damaged eye who always dressed in dark three-piece suits and was addicted in old age, when the author knew him, to endless pipes of Prince Albert tobacco and packages of Tums, had been a barber in Manhattan for many years, his shop, where he was called "Nick," located in the Chrysler Building—Mayor Jimmy Walker was one of his customers. His favorite chair at home was in the enclosed porch of the 145th Place house the family had moved into after leaving Corona in 1941.

Born March 7, 1866 in Italy, in the same village (Bella) as his wife's, he had originally come to America with his brother near the turn of the century, working as a shoemaker in Boston, then returned home to get married and return with his bride to New York, becoming a citizen on May 1, 1901 and possessor of a certificate from Albany declaring him a "Registered barber." His father had been a barber, which meant acting on occasion as a doctor.

46 "The Rest Is Silence"   The title of the poem, Hamlet's dying words, is inscribed in gold letters upon the author's black foot-stool tombstone in Cedar Lawn Cemetery, East Hampton, which has already been placed parallel to a twin for his wife, whose chosen epitaph is "And Time Itself A Wound That Cannot Be Healed."

47 After an artist's music-box life of women and whiskey (oiled by an indulgent wife), Dutch-born Willem de Kooning fell prey to Alzheimer's, but painted to the very end, which came in 1997 at the age of ninety-three in East Hampton—local rumors claimed a brush was tied to his hand in the last years. He had needed, like Picasso, women as subject and object, until mind and body began to split at their seams into dancing ribbons of color.

Amy Elisabeth Rothholz, the pretty and vivacious daughter of Paula Trachtman and Peter Rothholz and step-daughter of the author, died in New York Hospital on February 20, 1983 at age twenty-five. Enchantingly self-absorbed and intent upon becoming a star, she had graduated from Mount Holyoke College with a degree in theatre arts and was in pursuit of an acting career in

130

Manhattan. She is buried in East Hampton's Cedar Lawn Cemetery, along with her Yorkshire terrier Pudding and calico cat Miss Boo.

Here should be enrolled further crucial additions (or ego-wounds) to the author's *Doomsday Book*. On the morning of January 30, 1976, his mother Lena, several months after being released from the mental institution that had been her home for more than thirty years, dropped dead in the bathroom of the second-floor apartment (2S) in Emily Towers at the intersection of 150th Street and 35th Avenue shared by her sisters Emily and Edith since August 1, 1964. Lena's death certificate lists a coronary attack as the cause, brought on by "diabetes mellitus and chronic schizophrenia."

Edward Butscher, Sr. died several weeks later in Flushing Hospital, age sixty-six, while undergoing an operation for colon cancer, unaware that the cancer had spread throughout his body. Husband and wife were buried in separate Queens cemeteries with their respective families, Lena's funeral taking place during a whirling snowstorm. Edward's grey-suited corpse, a pint whiskey bottle tucked under his sleeve by his niece Lilian, was interred among the Butschers at St. John's Cemetery in Middle Village Section 21, Range M. Grave 306—another fistful of threes.

Short and beer-bellied with a receding hair line and thin-lipped Germanic face that women thought handsome, the author's father liked to believe that Irish blood sang in his veins, which later proved true, identifying with the tough, sometimes crooked, always heroic, characters played by James Cagney in many 1930's and 1940's films. In genealogical fact, his father John's lineage could be traced back to Alsace-Lorraine, and his mother Olga (*née* Carlen) was from a Swedish family that had been America for generations.

But later research by Carl Schlatter of Belfast, the author's first cousin once removed, indicates that the author's great-grandfather, Alphonse Butscher, had indeed married Mary Kingsley, an Irish woman, in 1881. The author's father is interred with his parents, his one-month-old daughter, and his sister Anna, who had died in her fifties from diabetes, after losing a leg to the disease. His modest stone on a small rise not far from busy Metropolitan Avenue reads:

BUTSCHER
John – Olga – Barbara
Anna Lypen
Edward Butscher
1911 - 1976

The poet under siege was forced to retire at age seventy by his college's mandatory retirement policy, a policy soon thereafter ruled illegal by the courts. Both his carefully crafted poetry and immaculate studio apartment affirm his gift for triumphing over a loveless self-imprisonment and its insidious closet death pressures, as does a kindly, if reserved, interest in his students, the author among them.

48 Georges Braque (1882-1963), wounded in World War I and master of the still life, co-pioneer with Picasso of Cubism, did a series of paintings of his ateylier in the 1950's. He is quoted in Jacques Damau's *Georges Braque* (New York: Barnes & Noble, 1963): "We shall never rest; the present is everlasting."

In his last sermon as dean of St. Paul's on February 25, 1631, an ailing, emaciated John Donne preached, "Wee have a winding sheete in our Mothers wombe, which growes with us from our conception."

For the author, Joseph Cornell (1903-1972) embodies the artistic triumph of an outer-borough provincial over Manhattan, refusing to abandon either Christian Science or his white-shingled, Dutch colonial house at 37-07 Utopia Parkway in Flushing—and the disappointed, naggingly critical mother and paralyzed brother who governed there. Constructed in his basement collected and bought materials, which swept in glasses, knobs, toys, machine parts, instruments and various print illustrations and movie stills, his "shadow box" assemblages elevate American middle-class yens for Hollywood fairy tales into resonating Surrealist dreams and nightmares.

The Museum of Modern Art mounted a generous exhibit of Cornell's shadow boxes from November 17, 1980 to January 20, 1981, which the author visited several times. Its curator, Kynaston McShine, also put together the catalogue for the show—*Joseph Cornell*, published by the Museum in 1880—which contains five informative essays on the artist's life and art as well. Plate VIII, "Untitled (Bebe Marie)," from the early 1940's is probably most relevant to the poem, imprisoning a female doll in full old-fashioned dress and hat behind a fence of branches, burnished with glitter, no thorns in sight.

Poet Simon Perchik's twin sister Esther died when he was twelve. The author reviewed *Hands Collected: The Books of Simon Perchik* (2000) for *Confrontation* (Spring/Summer 2002), 275-81, the first paragraph of which outlines their relationship: "I first met Simon Perchik some twenty years ago and was less than enthralled at the time by his 'smiling public man' persona. However, after reading two of his verse collections then available, *Twenty Years of Hands* (1966) and *Which Hand Holds the Brother* (1969), I found myself stunned into admiration, delighted at having encountered that rarest of native fauna: a powerful and unique poet. Friendship ensued and many chess games, much commiseration and playful shop talk, our poetry not always aligned but our views of the

poet's obsessive mission—to forge experience into lyric
vehicles of intense emotional and intellectual
apprehension—tacitly similar, as was our disgust at the
current state of literary affairs."

Later in the same review, when anatomizing a
specimen poem, the author observed, "References to
the war [Perchik was in the Air Force during World
War II, winning a hatful of medals in the flak-torn skies
over Europe], like the demagnetized colon, limn a
signature motif in the Perchik canon, allegorical in this
instance, as are the persistent images of hands and of a
dead twin, Esther, a mourned female self."

Poem #27 in Simon Perchik, *Who Can Touch
These Knots: New and Selected Poems* (Metuchen, NJ:
Scarecrow Press, 1985), p. 27, edited by Robert Peters,
selected and titled by the author at his friend's request,
is the purest articulation of the twin theme:

> Your breasts, little sister
> are drinking my perfume, are lips
> tasting steam, are clouds.
>
> Do you know, little sister,
> how much an ocean weighs
> when I sing from its peak?
>
> Little sister, in your tiny breasts
> are ships, are winds, are anchors
> soft and fine as silk :skies
> that only sailors really see.

The astronomer is doubtless Stephen W. Hawking,
victim since his twenties of the viciously progressive
Lou Gehrig's disease.

51 "Dust to Dust" Like W. B. Yeats, Rilke is often

identified with a tower, physically and metaphorically. In Rilke's case, the association is with the Chateau de Muzot, a twelfth-century tower in Switzerland, where in February of 1922 he completed the *Duino Elegies,* begun in 1912 at the Duino castle near Trieste, and *Sonnets to Orpheus* with a feverish burst of creative energy, the latter "written as a monument for Wera Ouckama Knoop." Crowned "that beautiful child" by the poet, Wera was the talented daughter of a friend, gifted in music and dance, who died shortly before her twentieth birthday. See Rainer Maria Rilke, *Sonnets to Orpheus*, trans. M. D. Herder Norton (New York: W. W. Norton & Co., 1942), p. 139.

Rilke died in 1926, four years after finishing his two masterful sequences. As translated by A. Poulin, Jr. in *Duino Elegies and The Sonnets to Orpheus* (Boston: Houghton Mifflin Co., 1976), supposedly addressed "To a friend of Wera's," the last lines of the final sonnet in the Second Series is the poet's monologue and monument to self, avatar for Orpheus, built, as were Shakespeare's sonnets, on the ego's arrogant assumption of posthumous fame:

> And if the earthly has forgotten
> you, say to the still earth: I flow
> To the rush water speak: I am.

55 "From Shore to Shining Shore"   The Yeats epigraph is from his "Two Songs for a Play" in the first volume of *The Collected Works of W. B. Yeats*, ed. Richard J. Finneran (New York: Scribner, 1997), pp. 216-17.

"Lucy" occupies a miracle place between the author's dotage and its recrudescent adolescence, where art is always stronger than truth. Her mixed heritage enables her to embody, in striking degrees, the careless beauty of Helen and sensual allure of Latin America, the male independence of Diana and nurturing passivity of an unmelancholy Marquez whore. In reality, she strides Corona streets like a replay of Olga's imperious strolls past whistles and air kisses decades earlier, serene in her knowledge of her body's electric magnetism and her intellectual destiny elsewhere, sharing the author's passion for psychology and literature and against mystical retreats from life's finitude.

The allusion to the Chinese Year of the Golden Dragon, which spanned from February 5, 2000 to January 2001, and to the burning of a church full of millennium sect members in Ghana, which was reported in the *Times* on March 17, 2000, puts the lie to the author's claim of 1999 as terminus year.

In the *Times* edition of August 6, 1999, Chester B. Crocker's "Death the Winner in Africa's Wars" lamented, "Africa remains torn by 11 wars involving 16 nations, and countless rebels and splinter movements in the Sudan. War and famine have killed about 2 million since 1955."

The summer of 1999, which the author passed comfortably in a porched, double-decked Victorian house by the sea, proved one of the hottest and driest on record. Monday, July 5th, after two days of 101 degree temperatures, Julia Johnson of Manhattan, age seventy-nine, became the city's first confirmed heat-related death, succumbing to hyperthermia. A health care worker found her unconscious, her body temperature reading 107 degrees.

By Saturday, July 10th, heat wave deaths in New York City reached twenty-seven, the ages of its victims ranging from twenty-seven to one hundred and two, though it was mostly the elderly who fell. This was the highest toll since seventy-six died in July of 1936 when the thermometer climbed to 106 degrees. A month earlier, a radio telescope established that the galaxy is fifteen percent younger than estimated by astronomer Edwin Powell Hubble (1889-1953), among the first to argue that the universe is expanding.

On Sunday in Florida, Leslie Alexis Marchand, age ninety-nine, died. Biographer and scholar, he had started the Lord Byron revival in 1947, which culminated in the three volumes of his *Byron: A Biography* ten years later, followed by twelve volumes of Byron's *Letters and Journals* (1968-1982), important additions to the author's library. Of Byron, Marchand justly remarked that he "spoke with the voice of the disillusioned modern world than with that of the nineteenth century."

July 26th in Boston, another academic close to the author's literary heart, Walter Jackson Bate, died at the age of eighty-one. The ideal image of a professor, distinguished-looking in tweedy jacket and vest, pipe at the ready, he had taught at Harvard for forty years. His *Burden of the English Poet* and solid biographies of Samuel Johnson and John Keats are among the author's treasures. Bate, in fact, had written three books on Keats, regarding the poet as "the best example of a really great writer of the last two centuries"—an honorable Romantic judgment.

Between these two passings, there was the annual sweep by Japanese police of the Aokigahara woods at the foot of Mount Fugi for the bodies of suicides. On Thursday, July 15th, they found the remains of seventy-five people, a 34.7 percent increase over the fifty-five found last year and the highest total since they began keeping records in 1947. Blame was placed on a recent economic recession. The year before, some ninety people a day had committed suicide in Japan.

137

There was also, nearer home, the flight of John
Kennedy, Jr. ("John John"), age thirty-eight, and his
wife Carolyn (*née* Bassette), age thirty-three, residents
of 20 North Moore Street in Tribeca for the three years
of their marriage, and Carolyn's sister Lauren, age
thirty-four, to Hyannis Port on Sunday, July 18th, for
the wedding of a Kennedy cousin. But the plane, with
John at the controls, never arrived. By the next day, the
Coast Guard was focusing on locating remains, hopes
for survivors already remote. The three bodies were
found Thursday in a shattered fuselage seven miles off
Martha's Vineyard.

For consolation, the author had the earlier
announcement by President Clinton, reported in the
Metro section of the *Times* (Thursday, July 1st), that the
"Flushing Line," the Number 7 IRT elevated line from
Main Street and Roosevelt Avenue into Times Square,
had been declared an Historic Trail, added to the list of
sixteen National Millennium Trails. This put the train
the author had been taking into the Oz of Manhattan
since childhood (often playing hooky to do so) in the
same illustrious company with the Alaskan Iditorad, the
Underground Railroad (New Orleans to Minnesota),
and the trail blazed by Lewis and Clark.

57 "Ode on Cunnilingus"   Ovid (Publius Ovidius
Naso), exiled to Tomis on the western coast of the
Black Sea by Emperor Augustus for his salacious
verses and possible involvement with the Emperor's
promiscuous granddaughter Julia, might have
appreciated this erotic, if duplicitously circular, swan
dive into salvation. He died in exile in 18 a. d. at the
age of seventy-five, his *Ars Amatoria* banned from
Rome's public libraries, but in the Epilogue to his
Metamorphoses—Peter Green's translation, from his
acute essay "Venus Clerke Ovyde" in *Essays in*

Antiquity (World: Cleveland, 1960), pp. 109-35—was as self-assured of immortality as Virgil: "My name will be cut deep and indelible into the future. /Wherever Roman dominion spreads over conquered lands./ Men's lips will speak me; and through all the ages/—If a poet may trust his prophetic gift—I will live."

59 "Treatise on Love and Beauty" The conjunction of the author's aunt, an Europa incarnation sentenced to a nursing home, is with W. H. Auden, whose post-World War II sojourn in New York's Greenwich Village was a lonely one and whose "Musee des Beaux Arts" limns an Old Master painting in which Icarus's fall into the sea goes unnoticed—see also "The Shield of Archilles" from 1952, both poems in his *Collected Poems*, edited by Edward Mendelson (New York: Random House, 1976), pp. 146-47 and 454-55. The "sinless son," however, is not Icarus, but Astyanax (Skanandrios), offspring of Hector and Andromache, hurled from the walls of Troy by the victorious Greeks, who feared his future vengeance upon them if he lived.

Leo Bersani's persuasive essays in his *A Future for Astyanax* (Boston: Little, Brown, 1976), keyed to Freud and Racine's *Andromaque*, follow the "stages in the *deconstruction of the self* [italics his] in modern literature" in terms of Astyanax's lost future (p. 5), although the most telling sentence for the author comes on page 10 in a focus on a writer's language: "[I]n the act of writing, the word itself seems to be experienced partly as an unsubstantial sign referring to meanings beyond itself, and partly as a sensuous object referring to nothing but its own shape, sound, and position in a design of numerous word-objects."

Born Constance Ockelman, Veronica Lake (1919-73) was a popular film star in the 1940's when she appeared in Preston Sturges's *Sullivan's Travels* (1941) and opposite Alan Ladd in two 1942 mystery melodramas,

This Gun for Hire and *The Glass Key*, based on a
Dashiel Hammet novel, the same year she made what is
considered her best film, *I Married a Witch*. But her
vamp appeal, as tartly judged by *The Oxford
Companion to Film*, ed. Liz-Anne Bawden (New York:
Oxford, 1976), p. 404, "owed less to talent than to her
intriguing hair-style which hid half her face."

60 Across from and in sight of Turkey, Chios is a
small Greek island where Homer supposedly once
taught in its Daskalopetra. There were a number of
brutal invasions over the centuries, conflated here into a
single onslaught.

63 "Engrossing Death"   The epigraph is the second
line of the two-line poem, "In a Station of the Metro,"
Pound offered as a hokku-like exemplar of his Imagist
aesthetic. Its first line establishes subterranean scene
and mood: "The apparition of these faces in the crowd."

Li Po (701-762) or Li Bai, son of a rich merchant
who spent most of his life drinking and traveling, is
considered one of China's greatest lyric poets. The
legend is that he drowned in the Yangtze River after
falling from a boat while drunkenly trying to embrace
the moon's image on the water. This might have
been fished from his popular "Drinking Alone under the
Moon." In more prosaic scripts, alcohol poisoning did
him in or, as some scholars suggest, mercury poisoning
from the Taoist longevity elixirs he was fond of.

On July 20, 1969, Neil Armstrong, commander of
the Apollo 11 mission, became the first man to walk on
the moon. A second lunar walk would occur in
November of the same year.

In the early hours of the morning, Aunt Olga
sometimes dangled a banner (realer than Hester's S)
from a bedroom window on the second floor of Emily

Towers, while her sister slept unaware. Cut in a ragged square from old white sheet, measuring 36" across and 38" high, the banner had two rectangles of cardboard from a Dr. Scholl's box sewn into each bottom corner for ballast. Its unfurling was accompanied by shouted tirades, which might have startled her neighbors but never resulted in calls to the police.

This was reported to the author by his younger brother years after the fact, when the aunts were in poor physical and mental shape. Printed clearly with a black magic marker, the banner's message was a condensed version of various thick letters Aunt Olga mailed to President Carter and the FBI. Reading the banner, the author was moved by his aunt's extraordinary strength and courage, never a sign of these frightening delusions in evidence during the many laugh-punctuated dinners and lengthy phone calls he shared with Aunt Ida and her over the years from the late 1970's into the mid 1990's:

> I Am A Mature, Sane, American Born
> Woman Who Has Lived In N. Y. Quietly
> All My Life, Worked In Offices As Secretary.
> I Am Being Brain Washed, Invaded In Mind
> And Body—With Tortures, Maiming—Ostracizing
> Fears, Threats And Forced Ideas, Allegiances, ETC
> By Unseen, Unknown To Me—Vicious Criminals.
> I, An Ordinary, Private Citizen.
> I Have written To The F. B. I., President Ford
> The U. S. Senate And State. And City Police and
> Now—
> You, the People—For Rescue For Me, this
> Victim—And My
> Family And their spouses And Descendants—My
> Relatives And
> Neighbors…I Have Not Told What Is Being Done

To Me To Either
Family Or Anyone—In Fear For Their Safety…
I Was Born And Raised In Queens, Lived for 18 Years in
Manhattan—until May 1974, and Since, Again In Flushing, N.Y.
I Plead For My Mind And Body's God-Given Freedom
And Privacy—And Yours—In this Bicentennial
Celebration of our Country's Freedom—usurped
From this Private, Individual American Woman.
One American Is <u>All Americans</u>.
All Humanity..
In Spite Of these Inflictions On My Mind And Body—I Am
Fighting For My Freedom—And I Am Sane
Rescue! Help! Stand By Me!
Help Me. –And You!

<u>E. O. C.</u>
(My Initials)

64 "November's Fabulous Island Girl" Born in 1916, Aunt Olga, the "baby" among Nana's eight surviving children, was the family beauty from the start, petted and spoiled by her siblings, maturing into a voluptuous, vividly attractive woman with a quick smile and warm disposition. Finding Flushing too suburban for a single young woman with creative ambitions and a taste for sophisticated café life, she moved into Manhattan in the late 1950's—her modest apartment in the East Nineties was within walking distance of her beloved Metropolitan Museum of Art.

Known as Emily by her city friends, Olga wrote a large number of popular songs and a full musical in the *Oklahoma* mode (now in the author's possession) without any commercial success, though convinced several of her tunes had been stolen by the agents she submitted them to.

She also became ensnared in a long-term affair with a married man that ended with the latter's death from cancer, which may have helped precipitate the 1974 nervous breakdown that compelled her return to Emily Towers and her sister Ida.

Booth Memorial Hospital, now part of New York Hospital, was originally a Salvation Army institution caring for unwed pregnant women. Situated at the corner of Main Street and Booth Avenue, several blocks from John Bowne High School where the author taught English for twenty-nine years, it was the place Olga was sent when her emphysema and related ills had worsened to a dangerous point—a necessary prelude to joining her sister Ida in a nursing home required by law and done with the assistance of a kind Indian doctor not overly concerned with legal nicities.

In particular, Olga was upset by the memory that Booth was where her sister Mabel had died on June 18, 1979. When being wheeled on a table to the elevator taking her downstairs for the transfer to the home, she insisted on having her imitation leopard-skin jacket draped over her shoulders, sitting erect and regal, staring straight ahead. This drew laughter and a smatter of applause from fellow patients and staff members, who had found her difficult to handle at times. Lugging a bag of clothes in her wake, the author mentally etched Hatshepsut sailing serenely down a croc-infested Nile.

Sunday, June 14th, in Brooklyn, Olga Maisonet, age forty-one, walked up behind a stranger carrying a bag of groceries and stabbed him to death with a kitchen knife. The victim was Guo-Xi-Li, age seventy-six, a retired factory worker living with his daughter and granddaughter, who had emigrated from China's Guangdong Province in 1990. His attacker had spent the last fifteen years in and out of hospitals because of her schizophrenia. Lutheran Medical Center had cut off Maisonet's medication a month ago, and on May 22nd she had been arrested on charges of stabbing a woman in the street but was released five days later by the court when the paper work was found not in order.

The *Times* of August 5th reported from Manila (A7) that a tropical storm, code-name Olga, had swept across the Korean Peninsula and the Phillippines, floods and landsides killing scores of people, forty-two in North Korea, sixty-three in South Korea, five in Thailand, and forty-four in the Phillippines. Earlier flooding of the Yangtze River had killed four hundred Chinese, leaving almost eight million more homeless.

65 Wracked by osteoporosis and advanced senility. Aunt Ida had been taken to the Cliffside Rehabilitation and Residential Health Care Center at 119-19 Graham Court in Flushing, perched not far from the edge of Flushing Bay and directly across from LaGuardia Airport. One of the few pleasures left her, apart from ingesting the bland food the institution served, was the sight of huge jets landing and taking off, especially at night, which invariably triggered child-like smiles.

66 "Crooked Man"   The author's grandfather kept a beautiful set of straight razors in an expensive-looking black leather case, which went to his son William when he died on July 12, 1951. Uncle William was the only one in the family still using a straight razor. The case gleamed in the author's memory like a jeweler's box designed for dueling pistols.

69 "Objects of Desire"   The epigraph is from "Peter Quince at the Clavier" in *The Collected Poems of Wallace Stevens* (New York: Alfred A. Knopf, 1965), p. 91. There were several Vietnamese girls the author knew from Bowne, socially, not romantically, family members who were short and trim and quite pretty, as well as adept at working the system in terms of garnering financial aid and utilizing programs for the disadvantaged. They did well at school, worked hard and seemed to run the men in their lives—their relatives in Vietnam had been on the American side during the war. Pleasant memories of them fused with recollections of news stories about pirates attacking refugee ships, raping and killing all on board.

Once confined to the Cliffside Care Center, Aunt Olga's banked rage and paranoia tended to increase as her sister Ida, always the power inside their dyad, deteriorated into a physically handicapped infant who had to be protected against a staff damned (unfairly) as incompetent and treacherous. But however harrowed by anxiety and a progressive disease, she retained a strong narcissistic concept of her feminine allure.

Consequently, the author's twice-weekly visits had to include bringing her what she felt were essential to her self-image and Ida's needs, all of which were ordered purchased at the Woolworth's on the corner of Main Street and Roosevelt Avenue. Near the end, she began phoning the author at five or six in the morning of visiting days to hector him about saving their apartment and dictate the list of items required. Olga's voice at this point was more croak than speech, rasped further by displaced fury at the author, who was cursed as a "bastard" or "dope" for his crimes.

75 "Freudian Man"   Sad to admit, the name of the girl who inspired this poem is no longer recalled by the author, which appears appropriate enough in light of her metaphorical enlargement here. Doubtless she has married a Florida doctor and is often decked in comely tennis whites.

145

In the penultimate version of *Eros Descending,* there were two short poems, sophomorically comic, that the author deleted here for reasons of taste. Their crude humor did offer a break from the weight of the rest and were true to the erotic architecture of the whole, but they lacked the metaphoric layerings of companion pieces. One was a single-line poem, "Casino," that insulted Native Americans, and the other, which would have appeared at this juncture in the sequence, was a parodical Attic hymn to the Blue God of Viagra.

It was footnoted as follows: If it ever existed, pseudo as a homeless pod, this bogus translation from the Ancient Greek would have been home in "Anonymous Inscriptions" section of *The Greek Anthology* in the Penguin collection edited by Peter Jay (1981), pp. 49-51. The author's favorite among these is last, "On Plato's Grave," as translated by William J. Phillips: "Ascepius cured the body: to make men whole./ Phoebus sent Plato, healer of the soul." But among the "Anonymous Epigrams," the most intriguing chillingly relevant is "Inscribed on a statue of Hermes" in Peter Jay's taut rendition:

> I was not—was born—was—
> and am not. That is all.
> Anything else you may say
> is a lie. I shall not be.

On Saturday, April 10th, linguist James D. McCawley, age sixty-one, collapsed from a heart attack on the campus of the University of Chicago, where he had taught for the last thirty years. In a field dominated by Naom Chomsky's generative grammar—a pursuit of language's "deep structure" via syntax and rules of sentence formation—McCawley and several colleagues developed what they called "generative semantics" in the late 1960's, which they viewed as a natural extension of Chomskyan grammar but which Chomsky, who had been one of McCawley's teachers at MIT, rejected. McCawley was born on May 30, 1938 (the author's birth year).

While in leisurely pursuit of a Ph. D. at NYU in the 1980's, the author had been required to take two linguistic classes. In one of these, much time was spent deconstructing sentences to diagram basement structures, which had the virtue of reinforcing the author's more impressionistic grasp of semantic absences, the tacit presence, for example, of negative and interrogative forms in every native speaker's simplest statement. Circles of analogies were not far behind, amplifying the linear march against death.

The *Times* September 23rd edition (A27) recorded the death from congestive heart failure of Thomas E. Atkins on September 15th. A private among the American forces that had recaptured Manila in the early months of 1945, he was in a perimeter foxhole with two other soldiers in the rocky terrain of northern Luzon when two companies of Japanese infantry came at them in the middle of the night. His two buddies were killed, and he was wounded in the hip, leg, and back, but he held off this and other attacks single-handedly, using his rifle and the rifles of his fallen friends when his failed.

Four hours later, at 7 a. m., all three guns jammed, Atkins was surrounded by the bodies of thirteen Japanese soldiers. He went to get a fourth rifle and killed another enemy behind the lines. Even when being carried away on a litter, he sat up and helped force yet another group of invaders to retreat. As Corporal Atkins, he received the Medal of Honor from President Harry S. Truman on October 12th, later leaving the service to become a farmer.

Labor Day, September 6th, an article in the *Times*

reported that the nation loses a thousand of its World War II veterans every day, and in 1998, 550,000 veterans passed away, mostly World War II.

78 "Summer Composition"   Fecund number, symmetrically satisfying and rich in Freudian threes, a Pythagorean triangle hoisting the cosmos and these petard poems, nestled like inverse question marks in a missionary reversal (reversion?) to a perverse sucking upon the world's whole hole of atomic self-completion before the ax of language splits it into Plato's burning, yearning halves. Perhaps unrelatedly, the author has been often amazed by a younger generation's attitude towards sex, which includes the provocative, if sane, notion that various forms of oral sex do not constitute actual sex, are instances of "fooling around." Bless their immature hearts—and riper other parts.

A *Times* story of June 24th reported from Pristina, Kosovo that three male Serbians, a professor and two workers, were found dead in a basement room of the local university, two of them with their hands tied. They had been shot in a classroom and their bodies dragged downstairs. Serbs of all kinds were streaming out of Kosovo as Albanian refugees returned. The night before, British military police had arrested a Serbian civilian suspected of participating in the murder of nearly fifty Albanians in the village of Slavinje.

In the Sunday *Times* of July 18th, another report out of Pristina cited a public inquiry's estimate that ten thousand Muslim Albanians had been massacred by Serbian forces during their campaign to drive all Albanians out of Kosovo, although the latter comprised almost ninety percent of the province's 2.2 million population. U. N. troops continued to unearth mass graves, and in late May an international war crimes tribunal in The Hague had indicted Yugoslav President

Slobodan Milosevic and four top aides for crimes against humanity.

80 "Mole Man"   The epigraph is from the opening of Emerson's brief "Forbearance," which is in *The Selected Writings of Ralph Waldo Emerson*, p. 772. Loosening his clerical collar, trousers rolled, brave Emerson in "The Rhodora" (p. 766), preached, "Beauty is its own excuse for being." The setting for this poem is the Amagansett house on Hand Lane that the author and his wife would sell in December of 1999, moving to East Hampton and abandoning the backyard graves of Godfrey and Mrs. Podge.

As reported in the Wednesday *Times* of July 28th (A10), the Congo wars, which have been raging for the last five years, also involved a continuing massacre of wildlife that bordered on animal genocide. At the Kaheezi-Biega National Park, the entrance was cluttered with forty elephant skulls and the remains of three adult and one baby gorillas. Park officials estimated that one hundred of the two hundred and fifty gorillas in the park's highlands had been killed since 1996, plus three hundred of the four hundred elephants alive before the war—elephant meat cost half the price of beef for the locals.

The poachers had automatic weapons, which they had acquired from the Hutus in the Rwandan Army who fled Rwanda in 1994 after the slaughter of some half a million Tutsi and moderate Hutus, the Park a base for Hutu rebels and Congolese guerillas. It was calculated that of the five thousand lowland gorillas in the Congo before the war, twenty-eight hundred remained, and the mountain gorilla was nearly extinct, only six hundred and fifty left in Rwanda.

An article in the Saturday *Times* of July 3rd, depicted an ancient burial rite among the Tibetans in which a

149

monk, for a fee, carries the naked corpse to a sacred clearing and walks round it, reciting a prayer, before meticulously cutting it into pieces, paring flesh from bone. He then sledge-hammers the bones into pieces. After which, vultures perched in surrounding trees sweep down and feast upon the remains. In an hour or so, little or is left. This process is called a "sky burial."

83 "Pin Up"   The pretty and petite subject went off to pursue college and perhaps marriage to her high school boyfriend, still unsure of her desires and worth. The epigraph are the last two lines of a Catullus poem addressing Lesbia's (Clodia's) pet sparrow, whom she is deliberately provoking until it bites her, seeking relief from love's deeper anguish in minor physical pain. Francis Warre Cornish's translation for the Loeb Classical Library (Boston: Harvard University, 1995), p. 3, strives for vulgate directness: "Ah, might I but play with you as she does,/ and lighten the gloomy cares of my heart!" James Michie's translation in his *The Poems of Catullus* (New York: Random House, 1969), p. 19, turns the lines into a triple to better reprise the Latin's simulation of the bird's hoppings and peckings: "…I wish I could play/ Silly games with you, too, to ease/ My worries and my miseries."

84 "Nymphs"   The epigraph is from "What My Lips Have Kissed,/ And Where And Why," one of a handful of memorable elegies Edna St. Vincent Millay wrote that boldly confront the inescapable winter of love and life without recourse to metaphysical security blankets. They are in *The Mentor Book of Modern American Poets*, edited by Oscar Williams and Edwin Honig (New York: WSP, 1962), a solid paperback anthology (despite the absence of T. S. Eliot), which the author used in several classes at John Bowne High School with limited success.

The attractive Millay threw herself into radical politics and the arms of many lovers before wisely settling down with a devoted businessman husband, exchanging Greenwich Village for the hills of New Hampshire.

"Nymphs" generally refer to minor female deities, beautiful and graceful maidens, who inhabit woods, mountains, rivers, and other natural forms. But they can also be the young of certain insect species who undergo an incomplete metamorphosis.

The Friday, August 13th edition of the *Times* cites a Human Rights Watch report on Jordon that claims eleven women were killed by relatives in that country since January, "honor crimes," because the women were guilty of sexual violations, even if they had been raped, their killers escaping with little or no punishment

86 "Dream in the Afternoon"   The two lines of the epigraph are from what is probably Hilda Doolittle's most famous (and most effective) Imagist poem, "The Garden," which is in Conrad Aiken's Modern Library anthology, *Twentieth Century American Poetry* (New York: Random House, 1944, 1963), p. 163.

90 "T. S. Eliot Does Zaheda's *Kama Sutra*"
The epigraph is from Melville's *Clarel: A Poem and Pilgrimage in the Holy Land*, published in 1876 but dealing with an 1856 journey.

As a child, "Zaheda" swam here from Bangladesh, via Poland, brief and feisty as a wet tomcat, managing to defy conservative Muslim parents and the gossip of their Flushing neighbors. Disappearing into Manhattan after sneaking out to attend her high school prom, she went on to secure a CPA license while working full

151

time and in spite of her distaste for the profession. Sexy enough to be a lit major, she has since pursued a Vet career, tending two contentious cats named Plato and Othello and expanding her vocabulary by reading *End of the Road* and *Tess of the D'Urbervilles.*

In the late 1950's, Joseph Cornell often used to breakfast on apple juice and French fries at Bickford's then in downtown Main Street, Flushing, scoping the adolescent girls ("teeners") drifting or hurrying past, many of them on their way to Flushing High School, where the author was then erratically in attendance. In between sightings, Cornell pored over a biography or made notes in his diary, mostly "reports of nubile teenagers," to cite his biographer, Deborah Solomon—see her *Utopia Parkway: The Life and Work of Joseph Cornell* (New York: Farrar, Straus and Giroux, 1997), pp. 151-52.

Alas, Cornell, who died at home on December 29, 1972 from heart failure, died a virgin, though fortunate enough in his sixties to have found a woman willing to provide him with an alternate route to Xanadu.

Wednesday, August 4th, the *Times* had an obituary for Rudy Burckbardt, a Manhattan photographer and film-maker who drowned himself on Sunday in a pond near his summer house in Searsmont, Maine. Born in Switzerland, Burckbardt had been a part of the New York School ferment, a neighbor of DeKooning and friend of Fairfield Porter. Over the course of several decades, he had made ninety films, mostly 16-millimeter and under thirty minutes. In the mid-1950's, always fascinated by artists and poets, he made four films with Cornell. A Manhattanite since 1935, he remained enamored of the city—"what I love about New York is that it just grew up wildly."

152

The professor was Konrad Gries, chairman of the Classics Department at Queens College, with whom the author took several Latin classes.

91 Prodicus (460-399 B. C.) was a respected Sophist who pops up like spastic Punch in Plato's Socratic Dialogues from time to time. Only fragments and summaries of his writings, specifically from *On Nature* and *On the Nature of Man*, have survived him, and he had to endure the surrender of Athens to Sparta in 404 B. C. He has been credited with originating the argument advanced by the Epicureans to the effect that death need not be feared because it is not a life experience or, as Prodicus reiterated to his wife late at night, "Dead is dead."

After Olga's death on March 12, 1996, Ida was given a new roommate, though she seemed unaware of her presence and of her sister's absence—"Olga," however, was one of the few words she still uttered. The author's mother-in-law, Leah Nelson (*née* Chertov), formerly Leah Trachtman, already well into her nineties, was a patient in the Weshampton Care Center on Long Island at the time.

In Hindu mythology, Kali is the great and terrible mother, wife of Siva, who is both phallus and bull, both a destroyer and a conqueror of death. He is also the third part of the triad (Trimurte) with Brahma and Vishnu. Besides her necklace of skulls and girdle of snakes, Kali used dead bodies for her earrings.

92 The Welles film is *Touch of Evil* (1958). Godfrey was an extremely smart, if neurotic, striped tabby with some Siamese blood in his veins. The first several months of his life as a kitten had been spent in the closet of a co-ed's dorm room at Williams College, which might explain the neurosis. He died at age seven from feline leukemia but achieved posthumous fame of

153

a sort as an allegorical antagonist in the author's (as yet unpublished) anti-epic epic, *And Thus Spake Godfrey.*

Seng-chao (374-414), Buddhist philosopher and author of *Seng-chao's Treatises,* lived at a time when Buddhism, as well as Taoism, its Chinese imitation, was the dominant intellectual force in China. He denied the apparent flux of reality, insisting that all things are actually at rest, immutably so, and even time but another illusion. Most appealing was his notion that non-existence is not entirely vacuous, things and non-things having some meaning, especially to the philosopher who can harmonize his mind with the supreme vacuity. The author imagines that he had no children and enjoyed sitting alone in a candle-lit cave, projecting animal figures on the wall with his long delicate fingers.

94 "Objet D'Art" The epigraph is from Aristotle's "Poetics," as translated in this case by Ingram Bywater in the Modern Library compilation, *Introduction To Aristotle,* edited by Richard McKeon (New York: Random House, 1847), p. 657

Its front page blaring that the death toll from a massive earthquake that hit Turkey on the 17th had passed seven thousand, the August 20th *Times* had an article dated August 19th about North Korea emerging from a prolonged drought that had killed two to three million people out of a population of twenty-four million. The number of dead from the Turkey quake would continue to rise until on September 1st it reached fourteen thousand—an aftershock on September 13th adding another seven. Three days prior, a quake had shaken Athens as well, killing a hundred and one, although, to the reflex relief of some, none of the ancient sites were damaged.

Ernest Pappas (*né* Pappakostas), the second son of Greek immigrants, whose father owned a delicatessen in Flushing, died on January 12, 1999 at the age of sixty-two. Though a few years older, he was a close friend of the author's since the latter's Flushing High days when he was working after school at Eisen's Luncheonette on the corner of 35th Avenue and 34th Road. A birth trauma had left the heavy-set, unathletic Pappas with a skewed eye and slight limp, though it did not affect his voluble affability and keen sense of humor. He and the author passed countless hours sipping beers together at either the Chestnut or the Shanon Bar & Grills down the block from Eisen's.

Pappas worked for many years at Pan American Airlines, rising to a low-level management position and putting together their overseas tours, writing the necessary brochures, which he enjoyed most. He lost his pension when Pan American collapsed and had to scramble in middle age to find work, learning about computers. He finally landed at a travel agency that specialized in Greek tours. He and the author met every few months for dinner, usually at one of the many Chinese restaurants in downtown Flushing, reminiscing about the old neighborhood and its eccentric characters, teasing one another about their opposing political views, and refighting the Giants vs. Dodgers baseball wars.

Ernestos was the name used by the Greek Orthodox priest at the funeral, but Ernie's modest tombstone, oddly lacking any reference to dead parents and planted near the outer rim of Flushing Cemetery (Plot 48 in Division B) reads:

PAPPAKOSTAS

Ernie   We Hardly Knew Ye

1936   Ernest   1999

The death of Ernest Pappas washed away a major chunk of the author's past.

97 "The Heart Shuts"   The title is the line that follows the two lines in the epigraph. They form the climax of "Contusion," which Plath wrote seven days before her suicide. See *The Collected Poems* (New York: Harper & Row, 1981), p. 222.

In Manhattan on Saturday, April 19th, at about 6:26 p. m., a woman jumped in front of a train at the Upper West Side station, disrupting service for an hour.

At 2:15 p. m. on Sunday, June 20th, a fifty-six-year-old woman, registered under a false name at the St. Moritz Hotel, began throwing money ($38, 000) and jewelry out of her twenty-ninth-floor room, causing a crowd to gather. The police broke down the door to discover her standing on the ledge outside the window. When they tried to coax her inside, she backed away and ended up hanging by her fingers. Minutes later, she either let go or lost her grip.

August 12th, a Naples, Florida resident named Philip Martinetti, age thirty-six, leaped off the Trump Plaza in Atlantic City to his death after losing a large sum of money in the casino. And the *Times* of September 21st (B6) reported that Victoria Gianis, age thirty-seven, of East Hampton jumped from the sixteenth floor of the Plaza Hotel, dying at 12:50 p. m.

A fuller obituary in the *East Hampton Star* of Thursday, September 23rd (I,2) profiled a successful young woman who had graduated with honors from Sarah Lawrence and went on to the London School of Economics and studying art in France. Victoria then worked for her father's investment banking firm. In 1989, she moved to East Hampton after being mugged in Central Park, becoming a riding instructor and horse trainer. She also sang in the Baptist Church Choir and with the Choral Society of the Hamptons, but had been hospitalized earlier in the year for depression.

98 "Every Third Thought"   The title is from Prospero's final speech in Act V of *The Tempest* as he muses upon the surrender of his art and his island: "And thence retire me to my Milan, where/ Every third thought shall be my grave." The epigraph is from Longfellow's "The Three Silences of Molinos," dedicated to John Greenleaf Whittier, which the author found in a large, leather-bound edition of *The Poetical Works Works of H. W. Longfellow* (Boston: Houghton Mifflin and Co., 1879), p. 832. The two-volume set was purchased for eight dollars in Boston at a going-out-of-business sale. Its second volume bears a later publication date and is thus not a real "first."

Edward Cutillo, age twenty-two, died at 4:30 a. m. on Sunday, October 6, 1929, at Flushing Hospital. A teacher and lover of books, he had apparently been closest to his younger sister Edith—the author has several books that he had given to her, Edith writing boldly on the first page of each, "Edward Cutillo to Edith Cutillo." One of these is *A Guide to the Works of Art in New York City,* a slender, hard-cover volume with photographs, edited and published by Florence N. Levy in New York, dated May, 1916.

Emma Olga Cutillo died on March 12, 1996 at the Park Hospital shortly after the author and his wife had departed for The Hague to take in a special Vermeer exhibit. Before leaving, the author had an unusually serene visit with his much weaker, softer aunt at the hospital. A day later, Olga carted Melville's "carpetbag of ego" across the Flushing border into eternity.

In Jericho, Long Island, according to a September 5th *Times* story (33), a family moved into a split-level house they had purchased for $455,000 from Ronald

Cohen and stumbled on a 55-gallon drum in a crawl space. They asked Cohen to remove it, but he refused, so they wrestled it to the curb for the sanitation men to pick up. However, they would not take it, fearing it might be toxic, and Cohen came with his real estate agent on September 2nd and pried open the lid, only to confront another container inside. When that was also forced open, it revealed the mummified body of a diminutive pregnant woman, along with a pocketbook of make-up and an imitation leopard-skin coat.

Other clues to the identity of the woman, who weighed only 59 pounds, were a locket that read, "Patrice, Love Uncle Phil," and a wedding band. A follow-up *Times* piece on September 13 (B5) reported that investigators were sent to Florida to question Howard B. Elkins, age seventy-one, the house's builder, who had sold it in 1984 and bought a retirement home. Friday, September 10th, the day after being questioned, Elkins left home and did not return. His son found his body in the garage of a friend's Boca Raton house—he had killed himself by shotgun.

The September 29th edition of the *Times* (B6) gave the woman in the barrel a name, Reyna Angelica Marroquin, and a background. She was a Salvadoran immigrant who had worked in a flower business owned by Elkins.

Barbara Ann Butscher, Lena's second child, born on February 25, 1936 when the family lived at 109-38 on 43rd Avenue in Corona, was rushed into Manhattan's Bellevue Hospital a month later. She died there at 8:30 a.m. on March 25th, her death certificate naming an "intestinal obstruction" as culprit, "cause unknown." During one of their many alcoholic Sunday afternoons together, watching either the Mets or the Jets on TV, the author's father had insisted that the doctor who came to the house to treat Barbara was drunk at the time and had not washed his hands.

Statistics from 1998 in the September 16th *Times* (B3) indicated that 6.8 percent of every thousand babies born in New York City that year died before their first birthday, below the national average of 7.2 percent and on the decline, except among blacks.

Aunt Ida was actually referring here to her sister Olga, not her niece.

99 Although buried by the Coppola-Migliore Funeral Home, located next to St. Leo's Catholic Church in Corona, as had been her parents almost half a century earlier, Olga's funeral mass was conducted at St. Andrew's on Northern Boulevard in Flushing. After which, the traditional procession to the cemetery, which consisted of two cars, a hearse and a limousine, passed Emily Towers and then drove to Corona via Roosevelt Avenue and cruised slowly along 43rd Avenue, where both the Cutillos and Butschers had once owned homes, on its way to Sunnyside's First Calvary Cemetery.

100 "Thirteen Ways of Looking at Aunt Ida"   Born in Manhattan on February 10, 1911, when the family was living at 237 East 43rd Street, two years before the move to Corona, Edith Cutillo was Jennie's fifth child, the only red head and the only offspring destined to attend college—Ida enjoyed telling the story of Nana giving her a nickel and an apple for her trips to Hunter College during the Depression (she graduated in 1932). Over the next decade, she worked as an accounting clerk at B. Altman's, a Social Investigator for the Emergency Relief Bureau, and then the Department of Welfare, before joining the American Red Cross in May of 1944, three years after the remaining family— their parents, Olga, and she—moved to Flushing.

Unwed and essentially the head of the family, Ida assumed the burden of caring for aged parents grown frail, and then for Ronald, John, and the author after her sister Lena was committed to the state hospital in Central Islip, where Ida alone visited her every fourth Sunday, lugging shopping-bags full of food and clothing on the bus, the Long Island Railroad car, and the cab required to reach her. At the Red Cross, she became a Casework Supervisor within three years and would, on two occasions, be sent to a distant part of the country to manage the relief effort after a tornado or other natural disaster had hit.

In September of 1956, Ida secured a year's leave to earn her M. S. degree at Columbia University's School of Social Work, which she did, but her career with the Red Cross ended May 16, 1966 when she resigned in protest over the glass ceiling that prevented her from advancing as less competent men were promoted ahead of her. It was a brave decision in view of her home situation, but she did find a job as casework supervisor at Elmhurst General Hospital and would stay there for eight years, officially retiring June 14, 1974 with a small pension.

Edith Cutillo died at 6:15 a.m. on November 6, 1999 in Flushing Hospital, where she had been admitted four days earlier, seventy-six years after her brother Edward died in the same hospital. Despite his atheism, the author arranged for a traditional Catholic burial, as he had for Olga, the mass again said at St. Andrew's. He was the sole mourner, although a sprinkling of strangers and neighbors from Emily Towers were scattered among the distant back pews, and an amiable, old-fashioned Irish priest gave a warm, if necessarily generic, eulogy.

Later, standing before the open grave in pretended prayer across the river from Manhattan's sun-smeared towers, the author remembered mistakenly calling home one evening while working at Eisen's and being stunned into a panic by Ida's "Hello," instantly transported into the familiar, warmly lit Ethan Allen parlor snug around her. Unable to talk, he gently hung up the phone.

Among the books the author inherited was his aunt's copy of the *Rubaiyat of Omar Khayyam*, translated by Edward Fitzgerald and illustrated by Edmund Dulac (New York: Doubleday, Doran and Co., 1933) in which poem XXVI arrows to the terrified heart of Eros in Fitzgerald's Georgian archaic and Gallic punny version of its *carpe diem* sermon:

> Oh, make the most of what we yet may spend,
> Before we too into the Dust descend;
> Dust into Dust, and under Dust, to lie.
> Sans Wine, sans Song, sans Singer, and
> —sans End!

103 "Half Mast"   Like Emily Dickinson, who regarded him as a dirty old man, Whitman's entire oeuvre is a campaign against death, and the epigraph lines, which come from his powerful Lincoln elegy, "When Lilacs Last in the Dooryard Bloom'd," is a poignant example of a major strategy in that campaign. See Walt Whitman, *Leaves of Grass*, a Mentor paperback, its pages now brown and brittle at their edges, that the author purchased for fifty cents (New York: New American Library, 1954), p. 269, after reading "O Captain, My Captain" in class.

On Wednesday, March 24th, Laurence D. Duke, a retired state judge, died in Georgia at the age of eighty-six. As a prosecutor in the early 1940's, he charged and convicted several Klan members with flogging a black man who was found frozen to death. When Governor Eugene Talmadge said he would grant them clemency,

Judge Duke confronted him at a public hearing with one of the whips used in the flogging, and the Governor backed down. The judge and his family were the targets of threats, at one point having to live in a cell.

Another judge, a Federal one in Alabama, Frank M. Johnson, Jr., died on July 23rd from pneumonia at age eighty. He had also been the subject of death threats, as well as ostracism and cross-burnings, after handing down crucial decisions re employment discrimination, affirmative action, the rights of mental patients to adequate care, and the protection of inmates from inhuman conditions. In 1965, he issued the order allowing Dr. Martin Luther King, Jr., to lead his historic 52 mile march from Selma to Montgomery to protest the denial of black voting rights.

Sandwiched between these illustrious losses like hot corned beef was the London suicide by rope at age fifty-eight of "Screaming Lord Sutch," Wednesday, June 16th. His body was discovered by his partner, Yvonne Elwood, who revealed that he was on drugs at the time for recurring bouts of depression. Head of the Official Monster Raving Loony Party, Lord Sutch had campaigned in forty losing elections under the banner: "Vote for insanity—you know it makes sense." The only son of a policeman killed in the Blitz, he had legally changed his name from "David" to "Lord" in the 1970's and never wed but did have a son with a former American model who once rode naked through town as Lady Godiva during an early campaign.

Friday night, June 2nd, Benjamin N. Smith, age twenty-one, drove through the suburbs of Chicago, shooting at people he deemed "mud people," which meant blacks, Asians, and orthodox Jews, killing Ricky Byrdsong, age forty-three. The next day, he fired at and wounded several blacks and Asians near the University of Illinois, where he had been a prelaw student and which he had left in February of 1998 before being expelled over beating up a girlfriend, possessing marijuana, and posting racist literature. Sunday, he returned to Bloomington and killed Won Joon Yoon, a twenty-six-year-old Korean-American exiting a Korean-Methodist church.

Son of a doctor, Smith had fallen under the sway of Matthew F. Hale, a recruiter for the World Church of the Creator. The *Times* of July 7th profiled the organization, which boasted a brightly tinted web page offering a coloring book of white-supremist symbols and a crossword puzzle full of racist clues. The 'church' recruited aggressively on campuses and used sophisticated marketing techniques, making it the largest and fastest growing hate group in the country. After shooting Yoon, Smith stole a van and was chased by the police for an hour until he crashed at about 11 p. m. and shot himself under the chin as the police closed in, dying soon after at a hospital.

The Saturday, August 7th edition of the *Times* (B7) carried an article by Diego Gambetta, "Primo Levi's Plunge: A Case Against Suicide," revisiting Levi's death in his hometown of Turin at age sixty-eight shortly after 10 a. m. on Saturday, April11, 1987, when he fell or jumped over the rail of the stairs next to the lift outside his third-floor apartment, hitting the stone floor of the lobby. Levi's hellish ten months in Auschwitz in 1944, about which he had written with understated eloquence in his first book, *If This Is a Man* (1959), came back in 1986 to pummel him with a cycle of depression.

May of that year he said in an interview, "If there is an Auschwitz, then there cannot be a God." A month later, he told another interviewer that he and his wife were trapped, he caring for a senile, bed-ridden mother of ninety-five, she in the same situation with a blind, ninety-one-year-old mother. A prostate operation exacerbated Levi's melancholia in March of 1987, when he was even finding it difficult to read and write, although taking anti-depressants. An April telephone conversation with Elie Wiesel ended with Levi telling his fellow victim of Auschwitz, "Too late," when the latter asked if he should visit.

In a recent biography, *Primo Levi: Tragedy of an Optimist* (New York: The Overlook Press, 1999), p. 400, Myriam Anissimov concludes, "those who are determined to reach a clear-cut verdict on his final action can find arguments on the side both of despair and of hope in the future." But at 10:20 a. m., April 11th, Levi called the city's chief rabbi—it was Passover eve—and told him, "I don't know how to go on. I can't stand this life any longer. My mother has cancer, and each time I look at her face, I remember the faces of the men lying dead on the planks of the bunks in Auschwitz."

On January 2, 1987, Levi had written his last poem, "Almanac," to notch the New Year—see his *Collected Poems* (London: Faber and Faber, 1992), translated by Brian Swann and Ruth Feldman, p. 49. It offers both a farewell and a prophecy:

> Very soon we will extend the desert
> Into the Amazon forests,
> Into the living hearts of our cities,
> Into our very hearts.

From Amsterdam on Wednesday, August 18th, the *Times* quoted the annual Greenpeace report, which predicted the Amazon rain forest would be wiped out in eighty years if multinational logging companies continued their deforestation at the current rate.

An item in the August 25th *Times* dealt with the well-preserved corpse of an Indian found in a crevice in the West Canada Glacier, where he had apparently fallen hundreds of years ago. Dubbed "the ice man," he was described in a follow up piece (September 29th) as wearing a hat and a cloak, probably a hunter who had lived around 1450 a. d. This evoked images of a man tumbling into a stony darkness far from home, trapped and conscious to the end, linking him, in the author's mind, to Ezra Pound's conceit that poetry began with a man alone in a forest talking to himself.

The same day, Father Mario Zicarelli, age seventy-eight, died in a Nyack, New York hospital from the complications of a stroke. Born in Rimini, Italy, and raised in Brooklyn, Zicarelli had studied theology in Rome but also Italian literature at Columbia University. Ordained in 1947, he served in the Belmont section of the Bronx for fourteen years and chaired a local school board, where he took the radical position that spending tax dollars on parochial schools violated the separation of church and state.

Despite protests, he was transferred to blue-collar Poughkeepsie but retained a defiant liberal bent and argued against his church's curbs on family planning and abortion, declaring in 1974, "I don't think the Catholic Church has any business meddling in sex."

The teasing reference here to communication between "Anselm" and the "Monk of Fools" parodies the philosophic conflict between Anselm or Anselmus, Archbishop of Canterbury from 1093 until his death in 1109, and Gaunilon, identified as "A Monk of Marmoutier." Despite his Augustian belief in *"Credo ut intelligan"* ("I believe in order to understand"), Anselm had foolishly propounded an ontological argument for the existence of God, positing in essence, "that that which nothing greater can be conceived" must exist, since man had conceived it, granting it ultimate existence in understanding *and* reality, hence making God's existence a necessary first truth.

To which, the monk countered that this meant one could argue that a perfect island exists, because if not, any existing island would be more perfect in our understanding, and thus a contradiction. Imagination, in other words, can dream extremes, but the existence of lesser models is never proof of their validity. See St. Anselm, *Proslogium; Monologium; An Appendix in Behalf of the Fool of Gaunilon; and Cur Deus Homo*, translated by Sidney Norton Deane (Chicago: The Open Court Publishing Co., 1910).

The author's friend, Anselm Parlatore, poet, psychopharmacologist and the most intelligent man he has ever encountered, is a smiling, mustached paradox of relentless intellectual probings and reactionary politics, who moved his family from posh Southampton to an island off the coast of Washington in the early 1990's. Before the move, he and the author had maintained a playful correspondence in which they thrashed out a variety of aesthetic and literary issues. One Parlatore letter proposed, "In the notion of fore-pleasure there lurks all manner of perverse, and ultimately the possibility of the polymorphous perverse, the possibility of a text that would delay, displace and deviate."

The linking of life with literature is a mocking reprise of a familiar essay question on the annual New York State English Regents' Examination inflicted every June upon the state's high school seniors.

107 Framed and mounted, the poster, "*Hommage á Dali,*" hung in the author's Briarwood apartment and was created to promote an exhibition of Dali's paintings at the *Musé De L'Athenée Genève* that ran from June 2 to September 30, 1970

The Sunday *Times* of September 5th reported that pesticide-spraying helicopters were being deployed over North Queens and the South Bronx after the deaths of an eighty-year-old man on August 31st and an eighty-seven-year-old woman on September 2nd from mosquito-borne encephalitis—at Flushing Hospital

Saturday, September 25th, the *Times*'s front page announced that the mosquito-borne illness that had by then killed three elderly people in New York City might not be the St. Louis encephalitis originally suspected, was instead the avian-carried West Nile virus. Dozens of birds had died in and around the Bronx Zoo over the summer but were not reported, although the author knew of them because a former student of his was working there.

In the September 30th *Times,* a story from San Francisco about an article in *Science* magazine summarized that butchered bones of six people (two adults, two teens, two children) discovered in a cave overlooking the Rhone Valley in France offer the most conclusive evidence to date that European Neanderthals, who lived anywhere from 35,000 to 125,000 years ago, were cannibals.

Perhaps the First Elegy in *The Duino Elegies*, as translated by Harry Behan (New York: Peter Pauper Press, 1957), p. 5, should be cited, beauty eviscerated as "nothing/ but the start of terror we can hardly bear." And yet, defying descent and what Yeats' archly condemned as "the brutality, the ill breeding, the barbarism of truth," Rilke's Third Elegy reminds and consoles us that "when we love,/ a sap older than memory rises in our arms."